D1236936

To June

By
Monomoy
Light

May your heart be lifted
by the light of Monomoy —

Norbert Cain

By Monomoy Light

Nature and Healing in an Island Sanctuary

North T. Cairn

NORTHEASTERN UNIVERSITY PRESS
BOSTON

Northeastern University Press

Illustrations by Patricia Cousins

Parts of this book originally appeared in the *Cape Cod Times.* A series of stories and columns about a 1993 sojourn at Monomoy Light was reprinted under the title "Monomoy: Paradise Found" in February 1994. Articles on subsequent summer stays appeared in 1994 and 1995. Occasional columns and stories about the island have been published since then. All these materials are available upon request from the newspaper's library.

Library of Congress Cataloging-in-Publication Data

Cairn, North T.
By Monomoy light : nature and healing in an island sanctuary / North T. Cairn.
p. cm.
Includes bibliographical references.
ISBN 1-55553-448-1 (cl : alk paper)
1. Cairn, North T. 2. Monomoy Wilderness (Mass.). I. Title.
CT275.C254 A3 2000
974.4′92–dc21 99-086660

Designed by Diane F. Levy

Composed in Bodoni by Wellington Graphics, South Boston, Massachusetts. Printed and bound by Quebecor Printing, Brattleboro, Vermont. The paper is Quebecor Liberty, an acid-free stock.

MANUFACTURED IN THE UNITED STATES OF AMERICA
04 03 02 01 00 5 4 3 2 1

To the men of Monomoy who helped the healing
Hillary LeClaire
Ed Moses
John Hay

Contents

The seat of the soul is where the inner world and the outer world meet. Where they overlap, it is in every point of the overlap.

—Novalis

Foreword

ORDINARILY, a book that combines sand dunes with lighthouses, some seagulls, and an old dory thrown in for the sake of local color will lure any reader eager to hear more about the "good ol' days." *By Monomoy Light,* however, does not fit the requirements. It is not easy summer reading; it takes attention. That means, in effect, that North Cairn's book amounts to a learning experience.

Now teaching nonfiction writing at Mount Holyoke College in South Hadley, Massachusetts, North was a feature writer and columnist for the *Cape Cod Times* and became its science editor. In the mid-1990s, she got the opportunity to spend part of three summers at South Monomoy Island, near the town of Chatham, living in a partially abandoned lighthouse-keeper's cottage, where she was to do her writing.

Monomoy, designated as a wilderness area, is not a wilderness in one sense. It is an outlying stretch of uninhabited dunes, to be sure, but it is constantly visited by birders, tourists, fishermen, clammers, and general itinerants temporarily stepping out of their boats. But, as she points out, it is wilderness enough, on an edge of time, or time-lessness, bordered by sandy shores littered with empty shells and the buried bones of marine animals, such as porpoises, dolphins, and whales.

It is comparatively easy for an inexperienced visitor to lose his sense of direction on the island, more defined by its sands than by any inland road or highway. It is a continual shape-changer, a landscape that is moved by the wind, sea levels, and tides. Its past, present, and future cannot be understood in static terms. Monomoy's shores and surfaces are always on the move, undone and reformed to correspond to a dynamic equilibrium beyond our grasp or control.

With devotion and determination, the author set about the lonely task of housekeeping but also began to equate herself with a newly exposed world around her. We find her taking up residence with a wild weather often hidden on the mainland and with animals seldom so close at hand. She and a solitary seal meet on the water's edge, and she sings to this ancient "silkie" with some success. She talks to a dying gull and becomes an admirer of the island's herd of deer. Personal encounters are balanced against research into scientific texts or information contributed by local biologists.

The measure of this book is not to be found in its facts and information, or the accuracy of its scientific data. What we call "wilderness" can never be fully understood in those terms. We are unable to know what we do not join. The poet Robinson Jeffers's phrase "Not man apart" describes it very well. Even the lawmakers of the nation cannot set wilderness areas aside and expect them to be known at the same time. *By Monomoy Light* is an eloquent statement, having to do with what transcends our own, often painful experience, though adjectives have nothing to do with it.

The sands have buried many a ship over a thousand years or more, and many shipwrecked souls can easily be lost to them, for they are indifferent to human survival. Ultimately, we are rescued by the sea beyond us, which makes no distinction between the lives that are in its keeping. The roots of beach grass keep us in place. We are one with gulls and terns, with deer and seals, alive in the still unnamed.

—JOHN HAY, *Brewster, Mass.*

Acknowledgments

I WAS NEVER NAVIGATING SOLO.

So many people contributed to making the original Monomoy project possible, and many, many more assisted in bringing this book to publication. It would be impossible to mention them all, but it is important to say that each added to the healing that Monomoy occasioned in my life.

Thankfully, you can never separate the love you hold for a place from the love you discovered for the people who shared it with you. In the end, all the loving merges—for the setting, the creatures, the colors in the sound of wind, the texture of rain and sand—for every living thing. That's the small, seismic change that shifts your whole life into a new territory, and when you cross that border, you might not know much, but you do know this: You've come home at last.

It was quite a revelation to a loner like me, as cut off from a sense of family and home as the island itself is from the mainland of Cape Cod—disconnected, and in time connected once more, by drifting back around, as it is in the nature of things to do. Of course, it all depends on trust—opening your heart to people who want to help, your eyes to a world full of benevolent joy and giving, and your mind to an awareness of the universe in which the center is not self but selfless.

Restoring a sense of connection to other people is one of the hallmarks of healing from trauma, so all those mentioned here and many others who offered their support and understanding should know that what they did helped, literally, to document the history of a place and save a life. My thanks cannot express the gratitude I feel:

To Hillary and Anne LeClaire, who participated so intimately in all the joys and obstacles to my living on Monomoy and writing about the island. I am grateful for the sense of family they offered so unselfishly to me.

To all the officials of the U.S. Fish and Wildlife Service, who opened the island and the lighthouse to me, and then gave so graciously of their time and expertise. Profound thanks to Ed Moses, former Monomoy refuge manager; Bud Oliveira, present Monomoy refuge manager; Sharon Ware, Monomoy refuge operations specialist; Stephanie Koch, staff biologist for Monomoy; Anne Hecht, endangered-species biologist; John Organ, wildlife program chief; and Paul O'Neil, biologist and technical assistance specialist.

To the staff of the Cape Cod Museum of Natural History, especially Susan Lindquist, former executive director, and the boards of trustees who allowed me to live at the lighthouse keeper's cottage for three summers and facilitated the grant for my sabbatical to research and write this book; also staff archaeologist Fred Dunford, who helped with information about Native Americans on the Cape.

To John and Kristi Hay, who offered every conceivable support—

from financial assistance and encouragement to summer housing—at each critical phase of this book. Without their help, this work never would have been completed.

To the editors and staff of the *Cape Cod Times,* who supported the project from beginning to end. Particular appreciation goes to managing editor Alicia Blaisdell-Bannon and the rest of the Features staff—who picked up the grueling, extra day-to-day work in my absence so that I could accomplish three summer stays at Monomoy Light and a sabbatical in 1995.

I am indebted to many people from the Massachusetts Audubon Society, but especially to field ornithologist Wayne Petersen; Robert Prescott, director of the Wellfleet Bay Wildlife Sanctuary; and Wallace and Priscilla Bailey of Chatham, for answering endless questions and providing background about the natural and human history of the island.

Likewise, geologist Robert Oldale of the U.S. Geological Survey contributed his expertise on the natural forces that have shaped and reshaped Monomoy.

Thanks to fitness specialist Suzanne Cummings-Giammarco, who worked to prepare me physically each year to make the endless portages of gear and books to the lighthouse. Without her encouragement and expertise, I would have been crippled by more than the priming of the hand pump at the keeper's quarters.

To Robert and Cozette Poirier, my gratitude for the welcome into the warm circle of their family and their camp on the Powder Hole. They provided rich detail about the history of Monomoy and what it means to survive, simply and beautifully, in a quiet place.

To research assistants Rebecca Mazur and Rachel Silver, my appreciation for aid in complicated computer maneuverings and tracking down elusive dates and facts. Many thanks to James and Harriet Weaver for their generous support in the final stages of the manuscript. Without them, I would not have finished.

To Anne Farrow, who read the work-in-progress, edited the final draft, and throughout gave such kind, wise counsel. She saw the far shore, even when I could not, and steered me to safe port.

To my editors at Northeastern University Press, former acquisitions editor Terri Teleen and editor-in-chief William Frohlich, my gratitude for believing in the project through its many incarnations.

To Emily Schatzow and Judith Herman, who were always there behind the scenes, as both ballast and rudder for the vessel of my evolving self, I am forever grateful.

And to all the people of Cape Cod who have helped to preserve the sense of paradise on Monomoy. May it remain so, always.

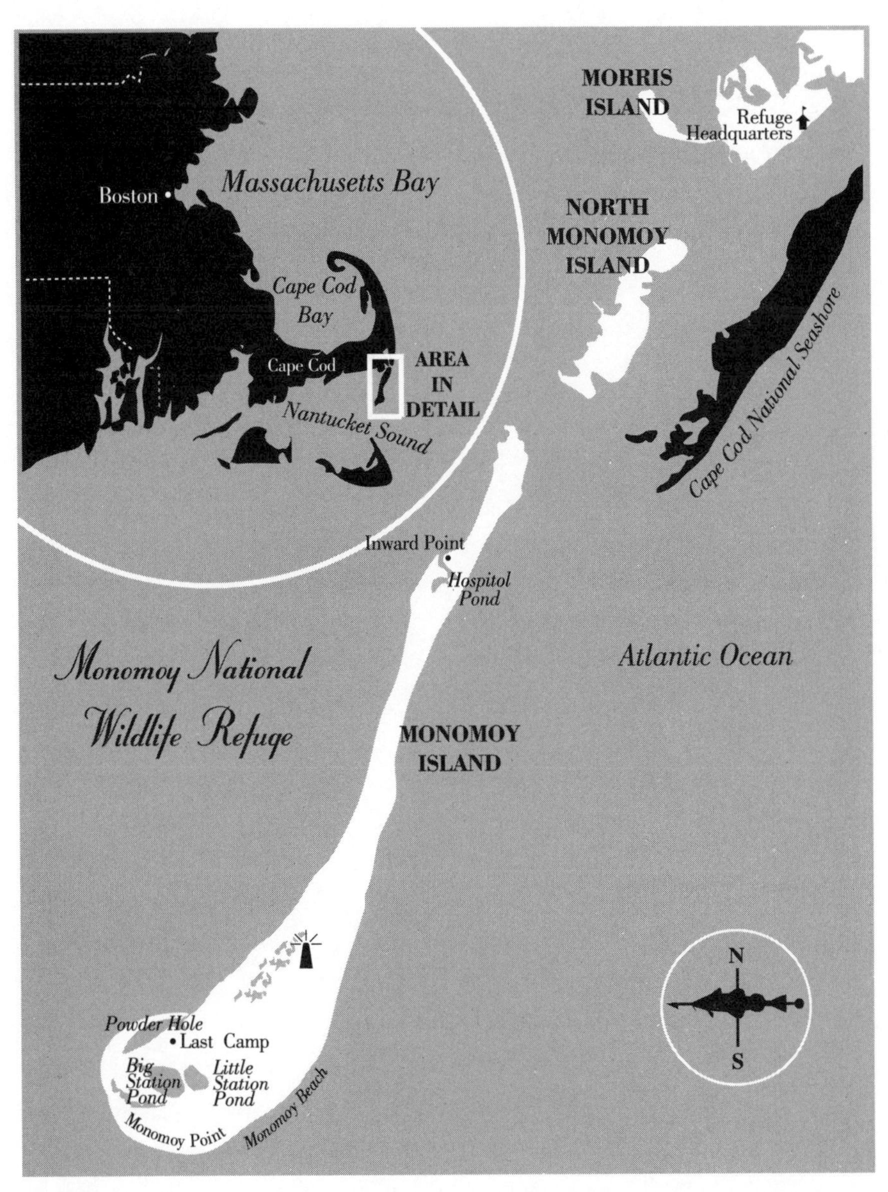

Map by Patricia Cousins

Prologue

Do not call it wilderness unless you are prepared to be lost and discovered there.

—JOHN HAY, *In The Company of Light*

ON NOVEMBER 26, 1993—the day after Thanksgiving—I stood at the bedside of my father, who lay asleep in a hospital bed in the family room of my childhood home in Chicago. This man, once an imposing figure, a well-to-do lawyer, formidable church leader, and authoritarian father, was now reduced by time and dissolution to something, someone, much smaller—a bedridden, defenseless old man, half out of his mind, crippled by the dementia of Alzheimer's or innumerable transient strokes, depending on the day's prevailing winds of medical opinion or inexact family reports.

I gazed at him for a long moment, taking in the transformed portrait of this man I had so feared while growing up. At last I lifted my hand over the nest of his pillow, barely resting my fingers on his bald head, his skin translucent and dappled as a gull's egg. I stroked his crown as one would a dreaming child's, marveling to feel in myself no expectation, no hope, no dread—only a sense of fate and destination. When, finally, he awakened, his pale blue eyes open and blinking like a nestling barely pipped from its egg, I halted my touch and began softly, slowly, to speak.

I had not seen him in ten years. For reasons that seem to me now as fleeting and predictable as tides, we had arrived, a decade earlier, at an impasse—a bitter rift over a long-troubled family history and its consequences. In the end—of that era, at least—I had taken matters into my own hands, surrendering to the instinct to survive. Migrating from the Midwest to the East Coast, I had fled my past, my family, and my home; changing my name and choosing the edge of the continent as a place to stay, if only for a time. I buried myself in hard work, harder loves, and the tiresome, if apparently necessary, rigors of self-scrutiny—searching to establish not blame but understanding, aiming not for justice but a bearable mercy.

I had not intended, in that beginning, when I left, to return home ever again. But I had not known then that family and its history cling, despite the long flight, like bits of shell and dried albumen, trailing a filmy membrane durable as DNA, connecting parent and offspring for life, indeed for all time.

Monomoy taught me that.

"HELLO, DAD, IT'S MARTHA," I started, in a whisper, trying to get my bearings with something simple—my birth name, no longer attached to me, but still the filament of sound that drifted back, down all the dark way, to childhood.

"Martha?" he echoed. "Martha. Are you staying?"

"No," I answered, a quick, hard word to sweep the solid world away. For a split second I was aware of looming over him, all the power and presence mine now, the universe of the past upended, turned in on itself. I saw the old chaos already reforming, reordering, the primal seas rising again, and falling; all that had gone before futile and suspended, like fossil remains. In an instant, in my mind, it seemed I watched the continents make their slow collide—the trivial lives, the insignificant years, whole species and epochs, entire, vanishing under the force of family.

"I'm just visiting for a little while," I heard myself say.

For perhaps three minutes we spoke then, the broken, private syllables of reconciliation and grief. I murmured forgiveness, he regret. All of it counted, and none of it mattered, for already we, and the long years, had flown—he toward the safer threshold of death, I to the uncertain border of life and risk.

I left that day, not knowing if he had been aware who had come to speak to him or what the words had signified. Perhaps I did not know myself. But in our encounter something had been released, some wayward evolution leading nowhere, the shared and faulty shame.

For the sum of all the aching years was not my history alone, nor his. It was the unfolding story of people and the places they choose for themselves, and in time become—a tale begun long before us both, unraveling and spinning, moving out ahead of our original, individual destinies like a wave breaking on infinity. It was a primer in healing brokenness and disconnection according to the instruction of the land, by living in wilderness and reclaiming the territory of simplicity, necessity, and joy.

As I walked the shores of Monomoy, those lessons became both grace and benediction. I had gone to the islands orphaned by devastation and will; there, I discovered the old relations, and returned, a daughter of the earth.

Break 1958

Plunge again,
Poor diver, among weeds and death! and bring
The pearl of brightness up. . . . It is we
Who, with our harmonies and discords, woven
Of myriad things forgotten and remembered,
Urge the vast twilight to immortal bloom.

—CONRAD AIKEN, *Preludes for Memnon*

I WAS SIX YEARS OLD, and Monomoy perhaps six thousand, when a nor'easter split the island from the mainland, leaving a narrow barrier spit about eight miles long in its wake. The place was fracturing always, sundered by the ocean surf under storm, wearing away, almost imperceptibly day to day, under the drag of the tides and time, worked by the perpetual winds. As it had been for six thousand years, as it is and will be.

South of the sand strip that is Monomoy, a coastal storm coiled into itself, centering, churning winds and water for hundreds of miles at the land's edge. The midnight skies of the last day of March dawned into the first morning of April 1958, while the nor'easter, spinning in its course and tumult, lashed the peninsula. The tides surged, rain and snow hurtled to the sands and disappeared into the sea. Men studied the weather with alarm, recording measurements, noting trends: winds rising to eighty-two miles per hour on Nantucket, just twenty-two miles from Monomoy Point; nearly five and a half inches of rain falling in twenty-four hours in Cummaquid, on the mainland. Nearby towns shuddered into darkness and seclusion as lights went out and phones failed. Windows shattered in the gusts, shingles split, and roofing gave way on hundreds of cottages along the coastline, as the nor'easter vented itself for two full days. "One of the worst storms in thirty years," the weather watchers told each other, tossing numbers into the wind, as though statistics might contain the storm.

But the gale managed its own turbulent harmonies, its wild, unimpeachable order. The waves spent themselves on the end of the northern beach of Monomoy, severing the bar of sand joining the barrier beach to the mainland, marking the end of a connection and a new, isolate era: the birth of an island.

And halfway across the continent, for the child I was, another storm, silent and secret, wore on. Betrayal descended like a gale, carrying me out, away from the known shore of trust, family, and home. By dawn, the breach was accomplished, parents disappeared by their own deceit, and I, cut off, wrested from safety.

Over Monomoy Island, the light of a first day broke.

An Extravagant Emptiness

I have come to a still, but not a deep center,

. . . .

My mind moves in more than one place
In a country half-land, half-water.

. . . .

What I love is near at hand,
Always, in earth and air.

—THEODORE ROETHKE, "THE FAR FIELD"

ENTER THE BONE YARD, the domain of the quick and the dead.

The words, like wind or a petitioner's fervent prayer, drifted, insensible, through my mind. I stood on an island with my back to the Atlantic Ocean, at the base of a small sandy gully, surrounded by tidal marsh, thickets of bayberry, and moors of compass grass. At my feet, and radiating outward in a rough circle perhaps twenty feet in diameter, were scores of shells—knobbed and channeled whelks, moon shells, razor and sea clams, mussels, and the occasional chipped scallop fan. For the moment, I was focused on the near sandscape, a spot on a small barrier island over which gulls criss-crossed, dropping their take of the sea onto the packed sand to jar open the shells and expose the vulnerable prey inside. I was scavenging the remains, poking through the unspectacular carnage for shells worth taking back to the mainland, when it occurred to me to take a look around and figure out where I was.

Easiest to say I was at the northern tip of the south island of

Monomoy, a wildlife refuge off the coast of Massachusetts, where the arm of Cape Cod bends at the elbow, its joint shared by the towns of Chatham and Orleans. Monomoy lies just south of there, at the confluence of Nantucket Sound and the Atlantic Ocean. Its name derives from a Wampanoag Indian word, which has been translated variously as "deep black," "mire," and "the sound of rushing water." In contemporary, colloquial usage, however, "Monomoy" refers to three islands, two of which form a barrier spit, reserved as a wilderness refuge for wildlife. The third island of the barrier complex is Morris Island, attached by bridge to mainland Cape Cod at Chatham and used as the administrative post from which the U.S. Department of the Interior's Fish and Wildlife Service oversees the refuge. Yet, if you were to consult an ordinary map, you might not find mention of any of these places. Frequently, even in atlases that show the narrow slip of land stretching south of Cape Cod and northeast of Nantucket Island, the islands' various names—"Monomoy," "Monomoy Island," "Monomoy National Wildlife Refuge"—will be omitted, perhaps because for nearly fifty years, the barrier-island complex has been devoted more to birds and animals than to humans.

It is one delicate edge of one great sea—from the mainland, a half-hour ride by boat to the south island. The destination: eternity. Here, in an island microcosm—a little tract of dunes, barrier beach, freshwater and brackish ponds, tidal flats, and salt marshes—something of wilderness still reigns. The place is given over to the governance of the natural cycles of seasons and migrations, daylight and dark night, full tide and ebb; and what is wild has ascendancy here, even in human law. Monomoy National Wildlife Refuge, the formal name given by Congress to the island sanctuary, was acquired by the federal government in 1944 for a wildlife sanctuary; and in 1970 lawmakers designated the island preserve as wilderness space "where the earth and its community of life are untrammeled by man, where man himself is a visitor and does not remain."

But the land itself needs no human legislation to declare its feral state. One sure sign is the natural indifference to individual death: In the wild, the living are too intent on survival to tarry over the dead. Thus, to walk any part of the islands is to stride among ruins and remains. Change, the first fact of life, and death, the final passage, are apparent everywhere: in the stripped and sun-scorched skeleton of a deer, fallen in the compass grass; in the mayhem of gulls that nested here and discovered death in the bayberry; in old beams of buildings once part of a port village, rotting now to slivers under the unforgiving sun, wind, salt, and restless sand.

Day after timeworn day, natural necessity rules here. Winds carry the driven birds to these shores for safe nesting in summer and rest in their long autumn flights, a full circle of faithfulness to imperatives no one wholly understands. Under the hot midsummer sun, deer rest in tall grass or amble without alarm through the moors to feed. From the sea and the Sound, the waters bear the eel grass in, and in its roll and snarl, a moon shell lies cradled, riding the foam, emptied of life. What once had substance and harbored a living thing—a predatory snail—is now turned to other uses, a skeleton suggesting something more: temporary shelter, perhaps, for a hermit crab; or if held in human hands, a tool for tuning to hidden currents, a symbol impressed upon the imagination, the symmetry of a thing coiled and complete—a single life, one pearled and sand-brushed shell.

The dunes create their own spirals, turning in upon themselves. The shores shrug under the lift and heave of the sea. The wave crests, like great hands wringing, twists rocks and pebbles, till all collide and fracture. In the hurl and thud of the breakers, everything once thought stable in its turn tumbles into tens of thousands of pieces, as grains of sand fall from boulders of granite—simultaneously, the unraveling of a continent, the creation of a world.

Listen then, and hear the voice of the wind, the whistling sand, and hushed surf. Heed the territorial imperatives—the reprimands of gulls, the declarations in the beach grass, and, in a sudden stillness that follows, the last long breath of the departed: *Enter the bone yard, the domain of the quick and the dead.*

FROM MY VANTAGE POINT, in that late summer—or early fall, adopting the birds' seasonal timing—of 1991, Monomoy looked like heaven, mostly untouched and certainly as unspoiled as it is possible for open land so close to millions of people to be. And there was this: Monomoy at least occupied a point on my internal map, a place in a physical landscape and in the cartography of the imagination, one that I could pinpoint, visit, describe, and study.

For one thing, it was small—about 2,750 acres, give or take the hundred or so acres of sand constantly eroded and redistributed by the sea. In scope and terrain it was manageable enough that human visitors in decent physical condition could walk the perimeter of the island in a day. A rigorous, healthy hike, to be sure, padding ten to fifteen miles through sand the entire way; but it was possible. And for the most part, there was only nature to take into account—a harmless wilderness made comprehensible through the lens of island isolation and the stresses that seclusion in nature entail. To travel there seemed somehow to entail little more than a frank and gentle human act, an admission of longing for a deserted place with the sea all around and the vast sky overhead. On Monomoy one could still pace an extravagant emptiness.

One could be quiet. And fathom the wind, the waves, the birds. And watch the day unfold and the creatures of the place working out their vital, tangled destinies. It was heady comfort to turn one's back on the civilization that lay only a channel's breadth away, look out over the waves of the Atlantic, and imagine the whole, far-flung world that lay beyond, safely out there, out of sight. It was tempting to forget that the insubstantial desert on which one stood was also a threshold inward, for the lay of the land within the self was not as simple to discern.

As I heard the words repeat, again, like a chant or a breaking wave—*Enter the bone yard, the domain of the quick and the dead*—I could not say for sure whether I was thinking about the space of sand on which I resided for the day or the animated dust I occupy for a lifetime, with a history of certain specters as companions.

Enter the bone yard, the domain of the quick and the dead.

It could have been a metaphor for the island. Then again, it might have meant me.

For some time I have wanted to tell the story of Monomoy and its slow accretion in my life; but, for years, I was not yet ready for the deep, internal archaeology the narrative might require, and I had not achieved the elemental perspective only time can grant. I could not disclose what lay at the core of either the place or the life. For a while, I did not know how to capture the island, its import, or its fate apart from the artifacts of my own past and evolving present—so indistinguishable had Monomoy become from my own constructs of meaning. Even now, I know that unearthing the island means uncovering buried significances in myself, for Monomoy overwhelmed, transformed, and redeemed me in ways I will be discerning for a lifetime.

But one must start somewhere.

For nearly fifteen years, I lived at one edge of the North American continent and called Cape Cod home. This peninsula of sand and kettle ponds, pitch pine forests, and dune beaches bearing on the broad Atlantic is Massachusetts' farthest reach into the sea. But when first I arrived, it was, to me, simply the point at which the land left off and there was no farther to run. It was not until years later that I realized that the uneasy sands of my own torn heart and interrupted destiny were being swept in the unforeseen direction of love and healing—by the place and its various worlds of sand and water, shadow and light. By that time, it was too late to leave without loss. By then, the place had become home. I had come to Cape Cod over a long overland route that carried me from the Midwest to Massachusetts, taking thirty years of life in the going. Looking back, it is hard to describe which tracks led where—family, schooling, career, loves. All the facts under certain lights become fictions, or at least stories kept alive only in the telling.

I was born and baptized Martha Ruth Mulder, the sixth of seven children in a quietly but deeply disrupted, secretive family. From the clearer point of the present, I can lift many memories to the light and turn them now, slowly and deliberately, casting half-lights like unreliable truths; but that is private work, reserved for the deep self and chosen intimates. Besides, everyone on the way to somewhere else has come from trouble of one sort or another, big or small. Details tend to obscure the point—people on the move are always fleeing a fate that no longer suits them for the unknown territory of change and fresh starts.

So it was that I, a child of the prairies and lakes, left Chicago for Michigan, and later, abandoned Michigan for Massachusetts, ostensibly to accomplish an education in theology, but primarily—or at least, additionally—to escape the circumstances of my birth and my family of origin. I left terrain that had been harsh frontier for my forebears two and three generations earlier and set out on my own hard and solitary pilgrimage, settling at last on Cape Cod.

For nearly twenty years, I have filled my days as a journalist, columnist, and now college professor, leaving the brown hours of evening and the brief, blue dawns free for wandering in nature. My self-imposed exile has taken me entirely beyond the reach of old connections to a new name and another family, born of the natural world.

In nature I found not only healing, but also instruction and, gratefully, self-transcendence. True things, it is said, come wrapped in clues that keep their meaning hidden, except to enlightened seekers. But whatever else the bliss of true understanding may be, it is elusive—usually hard-won and unexpected when it comes. I do not presume to guess that what I have arrived at is enlightenment; it is closer to Theodore Roethke's "still, but not a deep center." The more I have learned, the fewer words I have to describe faithfully what insights have greeted me like blessings along the way.

I only see that I came to rest—like one of the myriad migrating birds—on Cape Cod and stayed there for a decade and then some, taken in for a time by my own anguish and released at last by the beauty and grace of a safe harbor by the sea. I became enraptured by the particulars of the place—the habitats and their creatures; the slow trace of ordinary days, when weeds and wildflowers, insects and birds embodied the routines I came to count on.

During these years, I have spent a great deal of time in near-solitude and silence. I chose to live among—but not with—my own species. Almost alone, except for uncaring nature crowding all around, I spent untold hours with only the constant companionship of a dog—in the early years, a gentle Irish setter mix, and now, a golden retriever who not one moment too soon has achieved adulthood.

Still, for all this time, left to my own devices, I have enjoyed shaping a life that could hardly be called a lonely existence. I had work to do, inside and out; and I did it. Beyond that, I took the remaining sweet time—all my unhindered, uncounted hours—to acquaint myself with nature, and week after week to write in the corner of a newsprint page the little designs of my life.

After all this roving, through the days and weeks, seasons and cycles, I still cannot disclose the great ordering of my own small life or any other's. I know only that I departed from one place of land and water to arrive at another. The wayward years have not revealed to me the secrets of what commands the universe—or why. But I have come to see that I walk in the same light that awakens a box turtle, impelling her to trundle off, away from water and across roads and other human dangers, to nest as her ancestors have for millennia. My thoughts rise through the same air that the great blue heron consumes with wings, mounting up, out of the marsh; and my heart befriends the same currents the osprey navigates, crafting the ancient circles overhead. All these pilgrims—and their kin coursing in earth and sea—have become to me brother, sister, father, mother, and they announce my true home, the domain of the open heart.

It is tempting to dramatize the long walks all creatures make, and in prettying up the distances, to diminish what went into the years, and all that was lost. But in my own experience, the truth is more subtle, as slow as the growth of trees or the continent grinding down. Real transformation takes time and renders you as vulnerable as an amphibian taking on legs to replace what earlier had been the advantage of a tail.

I, too, know what it means to lose gills to gain lungs.

It is part of what it takes to move "in a country half-land, half-water." After a while, you begin, blessedly, to disappear, even from yourself, by fitting inconspicuously into the backdrop of sea and sand that encompasses everything on the edge of steady ground. In time, with luck, you glimpse what skills—and mercy—will be required of you, and what will be granted, so that you might start again, more freely, to live.

I HAD LIVED ON THE CAPE less than a year, barely accustomed to my new home and work, still in a vagabond state, when I decided to sever one of the last great bonds to my past: my birth name.

Even as I was sloughing off the old set of connections, I was casting about for the new. In seeking a chosen name, "North" came almost without effort. The sound was similar enough to my birth name to make the vocal shift insignificant, and the word itself—its literal meanings and its mythological resonance—seemed rich and full of mystery. I was drawn to the name as I had been pulled all my life by the far North Country—the white birch and pine forests of Minnesota's Boundary Waters and farther, the sub-Arctic regions of the Hudson Bay.

As a child I had imagined escaping from my parents' house to a score of safer places, protected from even the known intruders simply by means of an inhospitable climate. Other children dreamed of the adventure of space or expeditions to parts of Africa and South America that in the middle of the twentieth century still seemed remote and romantic. But I had other plans.

Displayed on my bedroom wall was a National Geographic map of the world, and all along the northern and southernmost extremities I placed colored tacks to indicate the places I wanted most to explore: the Poles, northern Canada, Hudson Bay; Greenland, Iceland, Norway,

Finland; Australia, New Zealand, and the Falklands. Year to year the relative merits of one destination or another might change, but the sites—and their far-flung locations, mostly to the north or to climates given to cold and snow—never varied.

In fact, I have never traveled beyond Ontario, have not yet made my way to the Earth's once-impenetrable frontiers. I have spent most of my life exploring the hinterlands of the hidden self, and when the time was right, I made the North—the mythological direction governing birth and death; the body and nature; growth, creativity, and silence—my own.

To discard one's birth and family name in favor of another is an uneasy affair at best, and to accomplish it, I looked to worlds beyond the human. I wanted the sounds of a place and the power of a totem to attend me into the new existence I was structuring for myself. One night, shortly after I had decided on a first name, the Scottish words "caird" and "cairn" came to me without prompting. I pulled down a dictionary and studied the definitions—"caird" from a Gaelic root, meaning a wandering tinker or gypsy; "cairn" with an ancient connotation of sepulcher or grave site, and in more contemporary usage, as a term indicating a pile of stones marking a change in direction. Either word and all the meanings would have worked, but in the end I chose "cairn," perhaps for no other reason than its less harsh sound.

Weeks passed before the loon—*Gavia immer*—flew into my consciousness; and, in time, I assumed a form of the bird's name, Tavia, and asked its blessing.

The morning after the black and white diving bird surfaced in my ephemeral, internal sight, I rose from sleep and crawled out of the sheets, my feet meeting the cold planks of the wood floor in what was then my coastal home. In the spare light of an autumn dawn, I drifted downstairs, and still groggy, leafed through a field guide to eastern birds. There, in the opening pages, were color plates devoted to the loons, some species specially ornamented with an occasional bright throat patch or head color. But the bird that had visited me was the common loon. "Large swimming birds with daggerlike bills, may dive from surface or submerge. . . . Sexes alike," the guide read. "Voice: In summer, falsetto wails, weird yodeling, maniacal quavering laughter; at night, a tremulous *ha-oo-oo.* In flight, a barking *kwuk.* Loons are usually silent in winter."

In the legends of North American indigenous peoples, the loon is variously credited with creating the Earth with mud or embodying the Creator's first incarnation, magically transforming from sound to shadow to the form of this most ancient of birds. Considered by ornithologists to be among the oldest of the avians, loons have a history stretching back ninety million years. Fossil evidence found in central North America in the nineteenth century demonstrates that birdlike creatures, strikingly similar to present-day loons, appeared about the same time as *Tyrannosaurus rex*—140 million years ago. And as *Homo sapiens* encountered the birds, tales of the loon's magical qualities were infused into oral traditions and depicted in prehistoric art. In native mythology, loons still occupy a place of reverence, being the bearer of shamans to the world of spirits.

That cold morning in the month of my birth, I knew my totem had come, and I draped the sound of its softened Latin name—*Tavia*—over me as a central and centering presence, reminding me at once of the dreams and hopes of the child I had been and the hidden depths of the fate that lay, unfulfilled, before me. For nearly fifteen years—hardly any duration at all compared to a bird older than recorded time—I have carried the loon, camouflaged in my name; and I have borne its spirit in my heart. The bird still skims through my mind with its odd postures and awkward flight. Sometimes silent, sometimes calling—it issues different songs in different times, in seasons and cycles governed by internal designs.

But mostly it moves in the fluid backdrop of phantasm, its low-slung form ghostly in the riffled waters. Now I know the loon so intimately that seeing it is like making a quick move in front of a lightly clouded glass and catching an indistinct, obscure reflection. The bird has surfaced in me during times of intense, fleeting joy and just before tragedy strikes. And occasionally, it has been the harbinger of the hardest passages, plunging out of sight just before word of a death reaches me. I ordinarily do not speak now of the times when the loon appears in my dreams, its song primordial and melancholy, full of a primitive loneliness and deep sorrow, its cry an unfathomable madness and healing solitude. You learn to live with your own wilderness, however it opens before you. Some things you keep to yourself. Certain revelations do not belong to any fragmented, personal history; they are history itself. They bear a wisdom so old that it carries you, and those who float for a time with you, over the waters of memory, on infinite lakes, whose source and completion are never found.

Flight brought me to the farthest shore of the continent, and here at the water's edge, I began to acquaint myself with the narratives of my own experience, to learn how much more there is to life than the struggle and survival of the body. Along the empty outer beaches of Cape Cod and its barrier spits, I have been instructed by the graces of nature about what it means to cherish a place one knows as home.

The lesson in its simplest form is this: To love a place is not so different from loving a person.

To love a place, you have to know it.

You have to leave yourself long enough to arrive somewhere beyond.

The philosopher who pointed out that love is where you live, and how, was on the right track. It is, after all, easy enough to love from a distance, in the abstract—to respect humankind, for example, or revere the Earth. It's the specifics that trip you up—or, under a different light, illuminate the finer points of what it means to love. Loving individual persons involves participation—in their particular existence and the shades of their temperaments. It means daily association—the unfolding toil, order, and joy of a life—sometimes disappointing and sometimes exhilarating.

Love requires allegiance and diligence, endurance and devotion. Places ask that of us, too. And what we first may be able to feel only for a place—a sun-strewn beach, a darkened thicket, a breaking wave—we may, with time, promise to other creatures and even ourselves. Until then, the land will cradle us.

DURING THE YEARS I visited the island on day-trip getaways, I envisioned Monomoy as a sanctuary. Later, during the summers I lived at the empty keeper's cottage by the extinguished light, I knew the place as a natural haven, a reclaimed island Eden, suspended outside of time and freed, somehow, from the ghosts of anyone's past or the erosions of personal history and hurt.

But, of course, it was never paradise, or hadn't been, anyway, for the long time humans had viewed it as a resource worth taking—a place to live at the edge of the sea, a limited territory with apparently limitless potential for hunting, fishing, and birding. Once humans directed burned-out barks and wooden ships, and later, dune buggies and skiffs, toward the shore, the era of the Island Garden on Monomoy ended.

I first visited Monomoy as the last decade of the century dawned, at the close of a fifteen-year period during which hostilities over the federal takeover of privately owned land in the refuge had, for the most part, ebbed. The abandoned lighthouse, keeper's quarters, and nearby storage shed had been refurbished in the late 1980s, just a few years before my arrival. Between 1970 and 1990, the number of "private" camps—owned by the government but leased for private use to the original land owners—had dwindled from thirteen to one. The last surviving shack rested on the shore of what once had been the island's sole commercial port—Whitewash Village, in the Powder Hole harbor of South Monomoy. And, in spite of storms, hurricanes, and human vandals working mischief elsewhere on the island, the two-room, tar-paper-and-wood cabin stubbornly endured until the close of the twentieth century. It was held, until 1999, under a life-use lease by its longtime owner, the gentle matriarch of a New Bedford, Massachusetts, family that had summered on Monomoy since the early 1930s. Diana Poirier, 94, died in late 1998, and as was stipulated by law, her cottage was scheduled to be dismantled by officials of the Fish and Wildlife

Service. Under provisions of the Wilderness Act, the Fish and Wildlife Service is authorized to remove the human remnants on the island and to restore a sense of untouched, wild lands.

By the 1990s, human visitors to the refuge already had become uncommon during much of the year. The early nineteenth-century buildings that stood locked and empty seemed to be all of a piece with the other skeletons exposed to the elements—birds, animals, ships' timbers, rust-flaked parts from motor vehicles, and concrete foundations for stations once used by the U.S. Life-Saving Service and the Coast Guard. And though the rotting remnants seemed bleak reminders of mortality and human transience, neither the symbols of decay nor the fossils of defeated human enterprise impressed me as tragic, or even finished. By then, I had lived for more than ten years near the ocean, and I knew that you cannot learn to love the sea if you are squeamish about death—or, for that matter, about the brutal contest for life. What goes on, goes on; and some things cease. So it was nothing startling to find that, on Monomoy, dry bones were as commonplace as twigs. The message scrawled in the unquiet sands became clear soon enough: Everything, in time, returns to dust—or here, to sand.

I came to know Monomoy during years when the refuge was, in one sense, at its best—seldom toured by people, remote and peaceful in comparison with the crowded New England seaboard stretching for two hundred miles north and south. Comparing Monomoy's Powder Hole with Manhattan, or Hospital Pond with Boston Harbor, the refuge's barrier beaches and dunes seemed as stark as desert.

But for all the emptiness of the place, Monomoy was—and is—more than touched by the human hand. True, nature is always working its demolition and reclamation projects in the sand, and the sea remains as ever that "well-excavated grave" that the poet Marianne Moore sensed it to be. Even so, humans have been vying for territory here for centuries, trying to gain the upper hand on the harmonious monotony of nature. The Monomoyicks, who belonged to the Algonquian family of nations, were the first to inhabit the stretch of land and among the last to see their identity as synonymous with the place, which they called *Munumuhkemoo,* an Algonquian word meaning "there is a mighty rush of water" or "the sound of rushing water." Later, European adventurers landed—often less by choice than by ill-luck in navigating the treacherous waters around the point and along the shores—and had their own names for a place so battered by the sea. The Vikings reportedly encountered Monomoy as part of early exploration, and later European adventurers, in attempting to chart the area, dubbed parts of it as "Point Fortune," "Batturier," "Tucker's Terror," "Point Care," and "Ungeluckige Haven." In all the names sailors and ships' captains bestowed on the island, the common theme was its unforgettable, dangerous character. In 1606, a ship carrying the explorer Samuel de Champlain broke a rudder on uncharted shoals, and the crew took refuge on Monomoy until repairs were finished and the ship could return to sea. From then on, the French explorer would describe the island's tip as Cape Batturier, "a bank on which the sea beats."

For all this, people kept haunting the shores and even tried periodically to wrest a living—or, at least, survival—from the fugitive isle. Early historical records show that in 1711 an early settler, William Eldredge, opened a tavern at Wreck Cove on the western, or Nantucket

Sound, side of Monomoy, "for the entertainment of sailors making a harbor in the vicinity." In 1729, the Irish immigrant ship *George and Ann* barely survived its four and a half months at sea, during which exhausted stores of food and water, disease, and, ultimately, mutiny claimed more than one hundred lives. The few survivors were rescued and taken to Wreck Cove, where they stayed the winter before pressing on to Ulster County in New York. In time, Chatham farmers used the expanse of inland Monomoy as pasture, and in the nineteenth century, a small harborside community, Whitewash Village, flourished for about seventy-five years along Nantucket Sound.

Most recently, the U.S. government has left its mark on Monomoy. The U.S. Life-Saving Service—later, the U.S. Coast Guard—maintained as many as four lifesaving stations from Chatham to Monomoy Point during the late nineteenth century. Then, during World War II, the federal government used the isolated strand as a strafing range. At last, as the war was nearing an end, the Department of Interior laid claim to the place, purchasing private land with the power of eminent domain. Life leases were granted to those whose dwellings still stood, but as time and storms devastated the camps, the owners were forced to leave so that the wilderness character of Monomoy could be restored.

People who knew the islands—or lived summers in dune shacks or shore camps—never really accepted the idea of relinquishing their interests or vacating the land. The hunting that had depleted large numbers of deer and birds was ended, for example, but even within the last twenty-five years, poaching has been a problem from time to time. And while the land has reverted to something akin to wilderness, the human presence is always there, in the shallows and along the shores.

Shellfishermen still comb the tidal flats, harvesting soft-shelled clams, and dig offshore for quahogs; professional and amateur ornithologists return each fall to the ponds and moors to glimpse migrating birds by the thousands; eco-tours continue to circle the refuge, looking for seals and other marine mammals; and boaters and sunbathers persist in using the beaches, even when their presence might disturb nesting or endangered birds.

Nevertheless, when I arrived in the mid-1990s for the first of three extended summer stays at the lighthouse on the southern of Monomoy's islands, the place seemed almost pristine, and I saw laid out before me a mostly unspoiled sanctuary of natural intricacy and balance. I did not have then the advantage—or cynicism—that can come with a long association with a place that both the government and the public find desirable. I did not realize that there had been trouble in paradise before and that soon enough there would be again. In 1996, a year after my last sojourn on the island, federal wildlife officials were embroiled in a conflict with animal rights advocates over the government's mandate to use necessary "management practices" to enhance the diversity of bird species in the refuge, which in this case meant destroying eggs in nests and poisoning the gull species that were overrunning the island. An estimated 2,500 birds—mostly great black-backed and some herring gulls—were killed that year with bread cubes that had been laced with poison and placed in nests. Several hundred birds flew to the mainland, to die in freshwater ponds in Chatham, where it was impossible to keep the public completely at bay. People all over Cape Cod, and eventually from around the country, were outraged by what they perceived to be the birds' suffering and by what they viewed as an intractable attitude on the part of officials of the Fish and Wildlife Service. In response to public protest, the use of poisons was suspended the next year; but 250 birds, including four black-crowned night herons, were shot, and eggs in nests were smashed. In 1998, federal officials attempted to make the island more compatible with nesting piping plovers and terns by eliminating the coyote—a relative newcomer to the island. Coyotes had extended their territory over Cape Cod during the previous decade, and in 1997, signs of a single coyote, and possibly a pair, were turning up on Monomoy. The coyotes presumably had swum from the mainland to the north island and eventually to South Monomoy. To prevent the animals from denning permanently on the island, one of the pair was shot and killed the next summer.

The intent of the federally mandated management program administered by the Fish and Wildlife Service had been to enable piping plovers and terns to nest on and inhabit Monomoy in greater numbers. That end was realized with dramatic success after only one year. But it came at a very high cost. The image of the agency had been tarnished in the public mind. Many people simply did not understand or accept the notion that human management was needed in a wilderness refuge, and they regarded the federal government's actions as interference, not intervention.

By the close of the twentieth century and the dawn of the twenty-first, Monomoy had been reconfigured once again as a landscape of bitter contention and a fragile, uncertain peace. To many, it had become, anew, a microcosm of the uneasy alliance between humans and the wilderness, the cynosure of our anxious motives in tinkering with the natural environment. Yet, contrary to what many people on both sides of the Monomoy controversy believed, it was not a simple matter to tell the culprits from the crusaders. Wildlife biologists found themselves ironically acceding to—indeed, endorsing—the culling of birds and animals they might elsewhere have been called upon to protect; and animal rights activists, bent on saving gulls, in time resorted to reprisals against Fish and Wildlife officials. Whatever innocence the island refuge might have embodied was eclipsed by squabbling and territorial jealousies—not of birds, but of the supposedly more sophisticated and evolved mammals: feuding humans.

Meanwhile, the natural forces that had shaped Monomoy continued to sculpt and resculpt the barrier islands. In the six millennia since its formation in the wake of the last ice age, Monomoy had been at the mercy of the battering surf and storms that sweep in from the Atlantic, and no human uproar would change that fact or its implications. Nature on its own guaranteed that Monomoy would never be a completely stable island environment. The erosion of land to the north carried sand down the arm of outer Cape Cod and eventually to Monomoy itself, continually reshaping the sand mass. Still, generation upon generation of migrating songbirds and shorebirds found the islands constant enough to serve as a regular stopover point on their southbound flights, thousands of miles long, to wintering sites in Central and South America. Razor clams and sandworms in the dense, dangerous intertidal zone went on with their existence, putting down feet like roots in the flats; and along the shore, beach grass thrived or failed, obeying the authority of the wind and sea, as the moving sand drove the island inexorably south and west.

Monomoy, which had been crafted by debris pressed ahead of great continental ice sheets and the rising sea level caused by their melting, rode out every change. For an island formed at about the time that early humans half a world away were discovering the advantages of farming, clustering into towns, and domesticating animals, the passage of time—with a decade more or less, a species more or less—was inconsequential. Who and what lived on Monomoy, how they came, and when they departed—all these were simply insignificant waves that meant nothing outside of the human calculation of time and distance, meaning and history.

Still, the last decade of the millennium was for me the beginning of ten years of wandering in a small wilderness, and it would change me for good. It is not that I found myself in the sparse wilderness of Monomoy, but rather that I lost myself there, in the intricate elegance and uncompromising energy of nature. To the island I brought, without thought, all my unanswered questions—what it means to be human, where I fit into the formula of nature and life, and how to understand meaning or affirm purpose, especially in a context of overwhelming suffering and loss. From my time apart I was expecting not answers but, perhaps, reprieve. On Monomoy I ceased asking why and turned instead to inquiries about what, how, where, and when natural mysteries revealed themselves. These questions focused my concentration on details that explained more about the communities of life around me and their connections to me than abstract pondering had ever done. Once my world was charged with an awareness evoked by hawks and voles, darners and mantises, beach pea and dusty miller, my human concerns receded to a more tolerable, even trivial, burden.

Now, with that exodus achieved at last, I am still. I sit in a simple, small, inland room, trying to make sense of an island and its influence—as place, as metaphor, as fact—in my own life and in the shared history with others of my time and species. It is almost ten years since I first approached Monomoy, five since my last summer sojourn there. I tell myself that space apart should give me a different outlook on things, and it does, since I am able now to rest in a candor and innocence that would not have been possible half a life or half a decade ago. I have come to know enough of brokenness and reconnection to trace the navigable bar of my own short, urgently human life, using shifting emotions for a space on which to stand and, finally, to report the stories. Only now can I give an account of the place—a barrier spit, severed from the mainland by the sea—and the narrative of my own internal territory, circumscribed by the loss, loneliness, and strength that a solitary pilgrimage brings.

The unreliable words themselves impede discovery. We hear the sounds, but the meanings evade us: mother, father, trust, tomorrow. Bird, animal, betrayal, love. Present, family, history, home. Still, as signals, these fragments may be the only hints we have, shards of a past, or evidence of a present, both so hard to hold.

And so it happened that during the early day trips and later, through the three, longer, summer stays on the island, all the language I brought with me was refined by Monomoy light and refired into syllables shaped by the silence of wilderness. In the swell of the evening tides, cradled in the arms of the wind, I heard the still, small voice of nature's law: *Enter the bone yard, the domain of the quick and the dead. Enter, and live.*

Thus, with no better compass to guide my passage, I looked to the island and followed its course, breaking the large tales into smaller pieces, sensing in the fragments some unseen, yet intuited, whole. In it resides the one natural, personal story worth remembering and retelling—the chronicle of survival: lives receding into narratives, ideas into words, thoughts into sounds, despair into sorrows, grief into longing—and, in time, into hope worth having.

A Refuge Reclaimed

With few exceptions, barrier beaches are examples of abused land healing itself.

—MICHAEL L. HOEL, *Land's Edge*

I DID NOT GO TO MONOMOY TO BE HEALED. That happened despite me.

I first encountered the island refuge the way most people do—not knowing what I was getting into but hauling out there, anyway, with the help of a friend. We went by boat for a day, to enjoy the escape from the routine of work, responsibility, traffic, and phones—all the clamoring demands of an existence that passes for life on the crowded mainland. My traveling partner and captain of the skiff was Hillary LeClaire, a born and bred Cape Codder who had taken up fishing after his retirement from the Marines, in which he had served as a pilot and risen to the rank of colonel—"spelled k-e-r-n-e-l," as he put it. Hillary, the husband of my novelist friend Anne LeClaire, and I had gotten to know each other in the way that people do when they meet in the margins of another, shared relationship and eventually form an attachment of their own. Hillary crossed to Monomoy frequently for shellfishing, and in his instinctively generous fashion, like an indulgent older brother, he let me stow away on his skiff when he headed out to dig clams in the tidal flats along the Nantucket Sound side of South Monomoy. Those, at least, became the excuses—his labor, my wanderlust—for our insatiable need to revisit the island whenever we could. The real reason was more likely the yearning for rest and, in the fresh air and uncluttered landscape, to find a little peace and quiet, the palliative that nature seems best able to arrange.

I had been working as a newspaper journalist for nearly fifteen years, specializing in nature and science writing for more than a decade, when, in the fall of 1991, I first joined Hillary for that thirty-minute crossing from Oyster River in Chatham through Nantucket Sound to South Monomoy. Partly because I was a nature writer and partly because I was on vacation for a few days and needed a retreat, Hillary had insisted that I come out to the island for a morning. He checked the tides and times for best navigation; I packed a change of clothes, binoculars, a field guide to eastern birds, a notebook, and pens and stowed them in a backpack by the front door. The next morning, Hillary called my cottage at 4:30 A.M. to let me know the weather and wind were right for a voyage.

"Just meet me at the house," he called over the abyss of half sleep. "We'll stop for coffee on the way to the river. See you at 5:30? Can you make it? I can't wait for you, so don't be late."

An hour later we pushed off.

I was converted to island life before we made it to South Monomoy that day. Perhaps it was the light, the company for coffee at daybreak, the sounds of the river awakening or paring down expectations and gear to a day's worth of time, but I couldn't get enough. I had made no plans for my days off, but by the time we put ashore at the refuge, I knew that I'd be tagging along every day Hillary and the weather would allow. We anchored near Hospital Pond at the northern end of South Monomoy, along the tidal flats that offered the best clamming. We went through the motions of being sensible, synchronizing our watches, agreeing on a time to meet back at the boat, then went our separate ways—I, to explore the island, and Hillary, into chest-deep water to scratch for quahogs.

One trip became two, then three, and finally four, as I accompanied him each day that week out to the island and back again. Slowly and without deliberation, I was being initiated into rituals of passage to a place that would come to figure large in my life. But it was still the beginning then, and everything seemed simple, uncomplicated, and new. The litany of tasks associated with the voyage out was always the same: Meet at the house or dock just after dawn, pick up coffee and muffins or doughnuts from the local fishermen's haunt, Larry's PX, a breakfast-and-lunch hangout on the road to the pier. Load the gear—clam rakes and rings, wire buckets, burlap bags, and plastic crates for steamers; waders or rubber boots for working the tidal flats; and a backup tank of gasoline for the trip home. Shove off and wend slowly down the river to the channel and the Sound, taking it easy so that the skiff creates no troublesome wake and there is time for coffee, banter, and the opening of the day.

Besides the two of us, there frequently were a few other fishermen milling around the shacks at the Oyster River pier to offer terse speculations about the weather or the day's wholesale prices on shellfish. A good day meant monosyllabic approval of the elements or the going price of clams; a bad day brought sullen, grim predictions bordering on existential despair. But with it all, there was always a healthy supply of teasing about one thing or another—Hillary's skill at keeping his scratch sites off Monomoy a secret, my hauling too much gear. And on the rare occasions when we failed to run into company at the shore, chances were good that we'd spot the men or their boats on the flats or hail their crossing as we navigated out to the island or back to Chatham.

When I think back, now nearly a decade later, to all the early trips to Monomoy, I still see all the characters there, hear their gibes and laughter, their boots scuffling along the dock, and their gear being dragged over the fill of broken oyster shells on the gravel road or tossed with a thud from the bed of a truck onto the pier. But as much as I remember the fishermen and the ceremony of the brief voyage, I most recall the pale colors of the mornings; the particular quality of near-silence on the river, save for a fish jumping, a puttering motor, the squeal of rubber boots against fiberglass, and our drowsy voices, amplified as they drifted over water. Recollecting those excursions, I can feel again the sway of uneasy footing on the launch raft and envision Hillary rowing out to his mooring, the rhythm of his pull on the oars regular and steady through the gray waters.

We seldom had other stowaways, though we were never short of companions, considering the birds. Cormorants hunched on sandbars, some with their wet wings outstretched for drying, huddled on the small tidal bars of sand, casting dark figures like crosses at first light. As the day came on strong, the fretful terns would wheel and dive just off the bow, hunting for fish in the waters of the Sound. And always, as we approached the island, the bother of the herring and black-backed gulls erupted. Squatting in the sand, they greeted us with a haughty *gah-gah-gah* or made an aggressive, low sweep just over our heads as we landed.

Later that fall Hillary decided I'd had enough of beachcombing and clamming, took an afternoon off, and hiked with me to Monomoy Light. He had anchored off an open section of beach on the Sound, and after we secured the boat, we set off together, expecting to intersect one of the few trails that lead to the lighthouse. We stumbled through poison ivy, got tangled in bayberry thickets, and tripped up by waist-high beach grass; but all the while, Hillary—who in college had been trained in wildlife management—tutored me on the finer points of the island's natural history, its flora and fauna, and assured me we'd hit the trail at any moment. Listening, I struggled along beside him, through the dense brush, half-wondering about our inland route. Suddenly, he stopped short and looked at me.

"You don't take direction from anyone, do you?" he said.

"What do you mean?" I countered, bewildered.

"Well, you've never been here before, have you? To the lighthouse, I mean."

I shook my head.

"I've told you I know where I'm going, but you just can't follow me," he went on, his voice filled not so much with rebuke as recognition. "I see you have to strike out and make your own little path there, even though I'm breaking one as we go. Look, North, I know you haven't had much luck trusting people, but really, you can follow me here. I know how to get us there."

"Okay, Hillary," I murmured. "Lead the way."

Five minutes later we stood, stranded in muck to midcalf, in a marsh a quarter of a mile from the light. The mud and water were getting deeper, the brier thicker, when Hillary turned again and glared at me.

"Don't start," he warned. "I can get us there, really. The path is here somewhere."

I raised an eyebrow. "Hey," I said. "I trust you."

"You fool," he muttered, and marched off.

THAT WAS THE FIRST of many disorienting—or reorienting—hikes across the island over the years. Getting lost, or at least losing one's bearings, is seductively simple on Monomoy, a landscape that changes quite dramatically from one year to the next, even one day to the next. Fog can cloud even the keenest sense of direction and keep one lost, wandering until the weather clears. Storms often reshape the landscape overnight, as fishermen who worked the shoreline know. Their remedy is to position a broken lobster buoy or large piece of driftwood—something specific but ordinary—in the sand as an inconspicuous marker of their own "good spot" for quahogging, just in case a recognizable shore shifts under the force of a storm and familiar landmarks vanish. And these are hardly empty fears or pointless remedies. Everything about the landscape—even its natural and human history—seems capricious, except for the barrier island's endurance over time. After all, one nor'easter had severed the then-peninsula of Monomoy from the mainland in 1958, creating the island; and twenty years later, a blizzard had split the island in two.

But even with the fickle weather and the wind- and water-drifted sands, I was seldom afraid of getting stranded or hurt on the island. I was often startled by the wilderness but hardly worried for my safety. It never occurred to me that a stern vigilance or inner discipline might be needed, not on those early trips out or on all the ones that followed in the rugged, kindly seasons that lay ahead. Two years later, when, for the first time I was allowed to live at the abandoned lighthouse, study the island habitats, and track the patterns of the birds' fall migration, I approached the island without alarm or defense. I never guessed that I was doing anything particularly risky that summer of 1993—or the two that followed—when I was permitted to occupy the forsaken keeper's cottage at the edge of the Atlantic. I did not foresee what Monomoy was to become in my life: the site and the occasion for an indisputable, interior continental drift that would shear away so much of the past, so imperceptibly at first, and leave behind the movable sands of deep and lasting change.

IT COMES AS NO SURPRISE to anyone who has spent a fair amount of time outdoors that nature has a healing effect on people in pain or distress. Part of it is fresh air and physical exercise. No matter what might be troubling you, if you take it outdoors and get moving, your attitude tends to change, even if nothing outside you has been transformed. Anybody, anywhere, is better off with some contact with nature.

But for people like me, people with lives disrupted by abuse in childhood, there's more to it. Trauma—sudden, devastating hurt—can happen once and leave its marring scar. But when it happens over and over, as was my plight, it becomes an imprint, which, unexamined and unchallenged, can be the trap you make of your own life long after the original damage has been done. Incest is nothing if not training; the seductions of the mind practice for a thousand furies.

These facts mean something very specific about healing, namely, that it takes time, compassion, and patience, and it must encompass every part of a person's life. Profound grieving is not just a sudden response to loss, a perspective that passes quickly so that you can get on to more important things. Nor is the restoration of your body and spirit, of your innocence and hope, something you can just think about and accomplish. The only way I know to recast years of a broken life is to draw from the animal instinct to survive, get to a safe place, and immerse yourself in patterns that can console you with harmlessness and beauty, constancy and love.

I remember, shortly after my first summer on Monomoy, talking to someone who had been on the island and was singularly unimpressed with it. "I don't get the appeal," he said. "There's nothing there."

"Exactly," I responded, thinking, *Yes, that's it. There's nothing there. All alone, nothing to keep track of, safe at last.*

But, of course, if that—a landscape with nothing in it—is the only place you can feel safe, you're in distinctly more trouble than you might have first guessed. The idea isn't just to escape or to live with such acute vigilance that you never know rest. You have to get away from the agony, for sure, but then the long migration begins in earnest—to find a destination where, once you've arrived, you can learn the endlessness of ordinary joys and sorrows, the refuge of being whole, the haven that home should be.

THE IDEA OF SPENDING WEEKS ALONE in the familiar wilds of Monomoy was a fantasy that in some ways took form as a lucky accident. I had been a stowaway on Hillary's shellfishing trips to the southern island many times between 1991 and 1993, so often, in fact, that in 1992, I purchased a nonresident fishing license from the town of Chatham to treat myself to clam digging on the tidal flats. My nonresident's license not only gave me access to the community's riverbeds and shorelines for limited clamming and scalloping, it also allowed me to dig along Monomoy's tidal flats.

At the time, I—like many people on Cape Cod—was working two jobs, augmenting my usual full-time employment as a newspaper writer with freelance writing and editing work. One job involved serving as editor for an annual journal for the Cape Cod Museum of Natural History, in Brewster. In addition to sponsoring educational outreach programs and nature study and research, the museum also serves as the federally appointed caretaker for the lighthouse complex on South Monomoy Island. That link proved crucial to my summers at the keeper's cottage.

I had been out to the island one weekend in the late spring of 1993 and had walked from the Nantucket shoreline around the edge of the Powder Hole. Eventually, I had made my way inland to the abandoned lighthouse, which as usual was boarded up to protect the building from the elements and from vandals. I had walked around the place and sat for a few minutes on the deck outside the keeper's cottage, thinking idly that it appeared to be a substantial structure, considering the light's remote location and age—nearly 150 years. It crossed my mind that it was too bad no one seemed to be using it. But I had come to the island for the day to do some birding, and my musings were interrupted by the sight of a marsh hawk circling the dunes leading to the ocean. I turned my attention to its flight, shrugged on my backpack, raised my binoculars to my eyes, and then wandered off to get a closer look, leaving behind my interest in the locked lighthouse.

I returned from the island that day rejuvenated and immediately was plunged back into the usual pace of day-to-day work. Midweek I had lunch with Susan Lindquist, who was then the director of the natural history museum, to discuss the editorial and photographic content of that year's journal. During our conversation, I mentioned my weekend hike around South Monomoy and my brief visit to the lighthouse. I suggested that the place seemed to be a valuable resource for natural history study, an ideal location to set up a naturalist's retreat for marine and wildlife biologists, writers, photographers, and artists. Each could spend a week or two in solitude at the keeper's cottage, with the understanding that the time there would be devoted to creating a scientific or impressionistic record of the place. The collected work, I suggested, might be published in the journal or, after a few years, compiled into a book as an island anthology.

Lindquist liked the idea but pointed out the hurdles to establishing a reliable schedule of fortnight stays. Every Monomoy plan depended on the vagaries of weather and wind, and travel had to be timed to the tides. These natural contingencies meant that no one could depend on the sort of ordinary routine that one would need in administering a program with a number of different participants. Even though unpredictability was one of Monomoy's principal attractions, it rendered any hope of organizing a consistent season of stays unlikely.

Weather and tides were the prelude to other obstacles, as well. Boat transport for people and their gear—most important, water, since there was no clean, fresh source at the light—was bound to be expensive, even under the best of conditions, and the mile-long portage of gear from the shore to the center of the island was likely to be too taxing for some people. Then, too, she indicated, the museum would have to win governmental permission for such a program and for each participant, as well as provide insurance coverage and a means of communication from the keeper's cottage to the mainland.

As she talked, it occurred to me that many of the barriers she outlined were problems for which newspapers have to make accommodations every day in covering news. I already had access to portable phones and surely was covered by insurance during work-related travel outside the office. Monomoy, I was convinced, was one of the Cape's last great, preserved natural places. A stay at the refuge, covered by a reporter, might qualify as a series of stories worth publishing. Lindquist agreed. But, she advised, if I were to go out to the island and truly get a feel for the place, I should plan to try to stay five days.

How about six weeks? I suggested.

In the end we agreed to a compromise of a monthlong stay, if approved by the newspaper and the museum's board of trustees. Then, she said, there remained the hurdle of permission from the federal government. That was likely to be the most insurmountable impediment of all.

THOUGH FREEING UP A REPORTER for a month was not easy for a midsize newspaper, the publisher and various editors at the *Cape Cod Times* approved the project. The newspaper provided a range of necessary supports, including a photographer to capture the visual experience of my island stay and portable phones for daily calls and to handle the transmission of copy each week. I submitted a proposal of stories that could be done, along with a plan for regular columns about the place, and together the editors and I devised a working plan for articles and deadlines.

Meanwhile, Lindquist broached the subject of having a journalist on the island for a month with Ed Moses, then the Monomoy refuge manager, who had a lifelong career with the U.S. Fish and Wildlife Service and had been affiliated with Monomoy since his youth. Lindquist confided to me that, in presenting the idea, she asked only that he not turn down the request without meeting with me and hearing the proposal.

I met with Moses and Sharon Ware, refuge operations specialist for the Fish and Wildlife Service at Morris Island, on a hot, bright day in midsummer. As it happened, I had brought my golden retriever with me in hopes of letting her go for a swim in one of the freshwater ponds of the Lower Cape before I headed back to work. It turned out Moses was a dog lover himself and owned a big, easygoing chocolate Labrador retriever. He told me later that he took the presence of a golden to be a sign of good character in me; and in the way that instincts and first impressions—intangibles over which one has little control—can make the critical difference in how decisions go, I think the dog, perhaps even more than I, won him over.

During that two-hour meeting, it became obvious that I would be cleared by the government to stay on Monomoy for a month during the peak of fall bird migration, from late August to late September. The only stipulations were that I had to produce material that could be used for educational purposes, as federal regulations about the use of the island facilities required; and I had to arrange for daily communication with someone on the mainland. Since fall migration coincides with hurricane season in the Northeast, I had to stay in touch regularly with the refuge, the museum, and the newspaper—just in case a bad storm loomed and I needed to leave the island for safer ground.

I drove back down the Cape that afternoon, dumbfounded by my good fortune, now and then looking over my shoulder at my oblivious golden retriever hanging out the side window, her ears blowing out like broken propellers from the sides of her head, her tongue lolling out of her grinning muzzle—as though nothing remarkable had just happened. But I knew better, and I could not believe it: I was going to live in a lighthouse, on an island where the only other inhabitants were birds and animals—and possibly a few members of the family that held claim to the last private camp on South Monomoy. But mostly, I would be all alone, in the most pristine and peaceful place I had ever known.

To flee to an island, or to unexpectedly land on one through chance of accidental calamity or sudden good luck, represents a fortuitous twist of fate that is a common dream. And because so many people today feel forced by finances or other family circumstances to lead what they experience as relentlessly hectic lives, the idea of a solitary island sojourn—especially in a relatively safe place, close to home—is the essence of escapist fantasy.

It was surely that for me. As a nature writer, I frequently had hoped to discover some place that could serve as my own humble Walden, a site removed from the press of ordinary life but near enough to civilization to visit people or to be visited. Without actually conjuring the details, I craved a place that was free of the complications or distractions of contemporary life, a landscape in which to measure myself.

And Monomoy was that space—in terrain, in experience, in time.

I was, perhaps, more suited than I ever imagined to blend, with a presumption of ease, into the backdrop of a barrier island in 1993. I was single, had no children to raise, was free of all attachments, especially of family, that might have made a monthlong absence impractical or worrisome. If I chose to disappear into the wilderness for weeks, I would not, to be blunt, be missed. The circumstances of my life mirrored, uncannily, the place I was to inhabit and get to know.

In a sense, I began to leave for Monomoy as soon as I had permission to go. Long before I landed on the island for my first summer stay, I was filling my leisure moments with planning, making lists, and storing up supplies. I tried to calculate what to eat and how best to prepare food that did not require a stove or open fire; what sources of light to use since there was no electricity or artificial light available; what clothes to bring to be suitably outfitted for every vicissitude of weather, from the desertlike, arid glare of the last weeks of summer to the biting, cold rains of the hurricane season and early fall. I pictured how I might deal with plumbing, because there was no running water for a shower or toilet, only a rusty old hand pump in the little kitchen at the lighthouse-keeper's cottage. For a while I even mulled over what, if anything, to take for personal protection, since there was no way to lock the most accessible door to the keeper's cottage.

As anyone who has lived for a time outdoors knows, planning is one of the most enjoyable parts of the experience, because the trials of the ordeal are still only imaginary. The more I envisioned the conditions of island life, however, the more I realized that the month away would be physically difficult and spiritually demanding. Island fantasy is one thing; actually being stranded, even by choice, for days or weeks on end, alone at a lighthouse, can be grueling, as stir-crazy keepers and their families learned long ago.

Nonetheless, I relished every second of the planning. I had done a fair amount of camping, and I had a reputation for making a home out of any campsite I occupied, so the notion of domesticating an abandoned lighthouse-keeper's cottage seemed more romantic than spartan to me. To prepare, I set aside one room of my house for provisions, organizing bedding (air mattress, sleeping bag, and muslin liners), food (dried stews and fruits, pasta, sauces, and dressings that required only a small amount of water or oil to complete, and coffee and tea), and clothing (layers made of fast-drying cloth, such as lightweight cottons and fleece, jeans and flannel shirts, as well as bathing suits for saltwater bathing, and hiking boots and sneakers with fiber uppers and thick soles for moving easily through sand). I gathered all the essential equipment (a propane stove, flashlights, emergency candles, polar sheets for warmth, a solar shower for comfort, and plenty of sunscreen, insect repellent, calamine lotion, and Fels Naphtha soap to treat—or better still, to prevent—the spread of poison ivy). I packed and re-packed everything to make the best use of space and to keep it dry; every collapsible item was rolled and stuffed into backpacks to make the portage easier.

Hillary helped with everything, monitoring my plans and gear from start to finish, bickering with me over rations and requirements for the stay. We made a couple of voyages to Monomoy in early August to cache camping gear, canned food, propane tanks, and seemingly endless gallons of water. But nothing prepared us for the labor of hauling in the small, heavy collection of reference books and field guides to aid my study of island ecology. Hillary grumbled over "the library" I had failed to mention, grousing every step of the portage and reminding me of the debt I was accruing for all his time, energy, and work to get me set up properly at the lighthouse. But in his weaker moments, he admitted that he was living my experience vicariously, enjoying the adventure of Monomoy while retaining all the benefits of leaving at the day's end to enjoy safe, comfortable nights at home, in an actual bed, with his wife.

Anne, meanwhile, graciously put up with all the intrusions into their life and its routines that my stay entailed—Hillary's absences while ferrying provisions to Monomoy, extra shopping trips for groceries, the endless wrangling and plotting about plans. She even agreed to be the person with whom I would check in daily; she would be available at 8 A.M. for a hasty call from the lighthouse. Over the course of the month, she took a few other calls, as well, answering my pleas for extra provisions, such as sugar and half-and-half for oatmeal and hot drinks; now and then, a dozen eggs, hard-boiled to make it across Nantucket Sound and through the hike in the sand; additional granola bars; and always, more coffee.

ONE DAY during the last week of August, we set out in Hillary's twenty-two-foot skiff from Oyster River in Chatham to make the final drop of gear. This time, I would stay.

The weather was ideal: eighty degrees, clear skies, wind from the south picking up, but well below the fifteen to twenty knots that could make the voyage rough going. The sun rose over the water, waves breaking the surface in pieces of light like shattered glass. On the wall of the boat's cuddy, our directions and destination were scribbled in mariner's code: From Outer Buoys: 225 degrees—6½ minutes. At cruise speed, 180 degrees—1½ minutes to trap 5 (Fish Weir). 190 degrees—7 minutes to trap 8. 190 degrees to beach.

We anchored off South Monomoy within sight of the lighthouse and made portage after portage to the keeper's cottage, until by early afternoon, all the gear for the month was piled on the deck.

Having arrived, we first faced the task of opening the cottage. The main door, alone, had four combination locks—the sequences of which

I carry to this day in my wallet as a memento of my time on the island. Getting settled inside the cottage was difficult for one person to accomplish alone, because once the door was opened, one still had to figure a way to detach, from the interior of the unlit cottage, the nuts and bolts that held the thick, heavy plywood panels on the outside of the windows and doors on the first floor. The upper floor of the house had plywood shutters that could easily be flung open from the small bedrooms. But on the main floor, the arrangement was more complicated and intentionally more secure, which made it a chore to open and air out the place. Once loosened from the inside, the large, cumbersome panels were likely to topple down onto the wooden deck before the person fumbling with the hardware indoors could get out to catch them. But Hillary and I worked together, and the job went smoothly and quickly, any sticking points oiled with the appropriate dose of nagging. When we finished, there was still enough time between tides to set up the propane stove and brew coffee.

Once the cottage was opened and the sunlight and the glare off the sand poured in, it was possible to take a look around. Several months' worth of mouse droppings dotted the shelves in the kitchen, and swallows' nests lined the outer sills of the upstairs windows. When I saw the scat—small as wild rice—mixed in with the few staples to be had—tea bags, coffee, dusty jars of spaghetti sauce and salsa—I set to work, cleaning. I found a large sponge someone had stashed under the sink years before, dusted off a dented can of cleanser left there, too, and started scraping the pellets into tiny piles and off into the trash. Meanwhile, Hillary worked beside me, installing a bright red hand pump from which was to come all my water for bathing and cooking—and, for me, a nearly lame shoulder after several days of working the handle.

It was a good introduction to the primitive conditions, and by the end of the first week of my stay, the accommodations actually struck me as quite comfortable. The cottage itself was huge for my needs—two floors above a forbidding earthen basement, which I avoided throughout my stay, on the theory that the rodent you don't see is one less intruder to worry about. In total, the main part of the cottage held six large rooms, not counting the dry bathroom upstairs, an alcove kitchen, a walk-in storage area downstairs, and a mudroom just inside the back door. There was no way I could fill all the available space with my belongings. I confined my living and working to one room on each floor, with the overflow of my library occupying a second large room on the first floor.

The room I used downstairs held two large tables—a '50s-vintage, wooden, oblong table with four sturdy chairs in the main eating area; a second, folding desk type of surface, shoved against one wall to form a long wide shelf for books, notebooks, and stacks of writing paper, which disappeared rapidly once I began handwriting copy for the newspaper. There was no power source for a laptop computer, and besides, every item for living or working was branded with blown sand within a day or two, which would make electronic equipment particularly vulnerable to breakdown.

Upstairs I erected a small, sky-blue dome tent for an indoor barrier against mice and bats, which came and went as they pleased. In it I spread a double layer of air mattresses and laid out a down mummy bag. From the tent's center poles, I hung a flashlight, and underneath, set up a small beach chair encircled by books. At night, when I took refuge in the tent, I felt as though I had embarked on a safari—as if trekking out to Monomoy was not distance enough to go. From the mesh screen windows of my tent, I could look out farther, through the windows of the cottage, past the wooden shutters, to a view of pines banked by dunes on one side and a vista of ponds and sedge flats out the other. Midway through my stay, the second full moon of a month was to shine; and the night before, its ivory orb of light spilling over the island was so bright that even after I had clicked off the flashlights, the shimmer seemed like a street lamp burning outdoors. I was glad for the glimpse of that lunar shine, because the next day I was temporarily called off the island because of the threat of high winds from a hurricane moving up the Atlantic Coast, and I missed the blue moon.

One evening early on in my sojourn, I climbed the steps to the second floor just before sunset and lowered the top half of a window looking out toward Monomoy Point. The bottom half was still supporting a barn swallow's nest, which I did not want to disturb, even though it was empty. By then, I was already caught up in my own insignificance and how much the interloper I was on the island. I stood, my elbows propped on the window frame, my chin in my hands, the barest, warm wind ruffling the beach plum outside, and watched the sun go down on the most peaceful day of my life. In the silence and unadorned purity of my days, I could sense—with my body and my spirit—that change and a nonhuman awareness of time were overtaking me. When the ragged current of everyday existence was swept away and the quiet came, I could discern the innocence of the natural world bearing down on me, and I knew, without knowing how, that nothing would ever again be quite the same.

THIS IS HOW THE DAYS WENT: SIMPLY.

And that, too, is how the healing came.

The sparrows and warblers would have been out, flying and feeding, for an hour by the time I rose, some time about 6 A.M., with the earliest light. I would lie in my sleeping bag and let sound prompt my first assessment of the day—which on Monomoy meant evaluating the weather. After living there for only a couple of weeks, a whole spectrum of distinctions about wind and the pounding of surf had been cast in my mind with corresponding judgments: too blustery to walk the beach; socked in by fog, too great a risk of getting lost; clear and calm, perfect for exploring, and so on.

My first act was to get water—whether that meant checking the priming of the pump, getting the primitive plumbing to drain, or just setting water to boil for coffee. That done, I toted a small chair out onto the deck, binoculars slung over my shoulder, and waited for the day to occur, while the songbirds wheeled in and the egrets stared into the shallow waters of the pond.

I belonged to the island now.

I'd taken off my jewelry, pruned my wardrobe to jeans, flannel and T-shirts, sneakers for indoors, and knee-high rubber boots for the out-of-doors. Though boots and jeans were not the attire for keeping cool in the sun, they did offer the advantage of blocking ticks and poison ivy oils. It seemed a long walk from the light to anywhere, though the actual measurements belied the sense of distance—the Atlantic nearly a hard half mile off, up and down through dunes, the same to Nantucket Sound. Everywhere felt farther away when I was plodding in boots a couple of sizes too big, through grass growing waist-high or in minithickets of bayberry, cattails, and sedge. Everything conspired to slow a person down.

The downshift almost immediately stripped my internal gears, habituated to run high on ordinary days at home. On the island, I was left to idle in the landscape, to watch willets and larks during the daylight hours, or to stop to admire the color and camouflage of a Fowler's toad against the sand. I took time to consider the effort of everyday tasks, weighing whether to scale the dunes, even if it meant a substantial reward, the chance to bathe in the sea. I forgot the details of negotiating life on the mainland. In the world of Monomoy, there were no cars (though every now and then I stumbled across some relics of earlier models left to rust in the sun and spray). No phones rang. There were no appliances—no dishwashers or washer/dryers, no VCRs or power mowers. I waited to think things through before acting. One thing about wilderness, you conserve your energy for the things that count—like making it to shelter or finding food and water—the routine matters of survival.

Once settled in at the keeper's cottage, I found no headaches had accompanied me there; I breathed easily and was not afraid. I used a solar shower and ate lightly. My right shoulder still threatened to go lame on me from the odd angle I had to strike to get leverage on the pump handle, but I adjusted. The considerations of moment on Monomoy ran to such issues as the merit of hand-washing a pair of jeans relative to the amount of pain to pump three buckets of water, or whether rice was worth eating if the cooking consumed twenty minutes of propane and necessitated lugging in another tank from the shore. Spend a few days ruminating on down-to-earth matters such as these, and you discover something. You find out what it really takes to live—how much lies right at your feet to be enjoyed, and how little there is to fear.

Early on, before I had left the mainland, as word got out that I would be "stranded" on the island, more or less in solitude for a month, I got one consistent bit of feedback: People were worried about me, being out there all alone. Even after I returned, acquaintances and even strangers would see me in my usual life—at work, grocery shopping, having lunch, picking up videos—and would stop to ask how I had fared. One question was inevitable: "Weren't you afraid out there, all by yourself?"

In fact, the question of my safety—because I am a woman—was raised several times by my editors, Fish and Wildlife officials, and the natural history museum staff. In planning my sojourn at the keeper's cottage, I never entertained the thought that I would face any more danger than I do on the average day in the world we think of as "civilization." If anything, I thought the risks would probably be less on the island than on the mainland. I knew, of course, that if I got into trouble, help would be farther away than ordinarily. But I live alone: Help—other than what I can give myself—is never at my fingertips. This level of peril is something I have accepted as part of the lifestyle I have chosen.

As my plans came to fruition, however, the level and intensity of alarm I encountered in people who care about me—and even from people who don't know me—got me feeling uneasy and reckless. I began to wonder why I wasn't afraid, started to feel foolhardy for being confident that I could take care of myself.

By the time I actually got to Monomoy, I was half-expecting trouble—and this is no disposition to be in when you are embarking on weeks of solitude. I remember the first day, when I was finally all alone, unpacking the tools that might come in handy—hammer, wrench, folding shovel, ax, and the like. As I stored them in a downstairs closet, I casually mused, in the detachment of horror-movie footage, how ironic it would be if an intruder used these objects as weapons against me.

And then, as idly as the anxiety had come, it passed. I looked out the unscreened windows into the dunes bleached by the white sun, listened to the wind, and calculated a distant sound to be surf. I watched the barn swallows navigate in air and took in the luminous grasses tossed in the gusts—and suddenly I was caught up by something that cannot sustain fear.

You could call it love. For the place itself, for the pristine sands spreading out in every direction, for an empty landscape with nothing to hide or fight.

Since that moment, when the fear fell away, I have thought a lot about the well-intentioned worry that went into my trip. Behind the fear, it seems, were beliefs worth examining: that anyone alone is at risk; that a woman alone is in real peril; that it is a dangerous endeavor, requiring real nerve, to head into the wilderness, even a manageable wilderness so close to home.

All these things may, in a way, be true. But I think the disposition to fear—beyond the reflexive, animal instinct that aids physical survival—is a choice; which is to say, there are other alternatives, different ways to feel. I prefer the sense of adventure, the exhilaration of the challenge, the gratitude for untainted land that touches the heart. And, as a woman, I believe it is important for me to seize the right to experience these things, alone if I wish.

One night during my first stay at the lighthouse, I returned by boat to the mainland just long enough to restock my food supplies, as light keepers had done throughout the nineteenth century. While there, I stopped to use a computer to file copy for a story that was running longer than I had first anticipated. Back at home that evening, I needed to get dinner, and the thought occurred to me that I had nothing in the house to eat. In my absence, though, four tomatoes and two lemon cucumbers had ripened in the parched, abandoned garden in the yard. After dusk, because all my flashlights were on Monomoy, I used a candle to locate whatever was still edible in the dark plot. I checked for chard and found gladiolas. Then I spotted the produce and took it gladly. Indoors, I dug into the household staples, found a half cup of rice, and, with the backyard harvest, created a feast.

There was a time when I would have felt deprived by such a meal, but I had been living in the lush desert of Monomoy. By now, I knew what it truly meant to be fed.

Living simply and in solitude is difficult, admittedly, since it strips you of distraction and defense. You find out the gravest danger you face—always—is yourself, and that you are your own way out of trouble, the doorway to your own hard-sought freedom. These are truths not everyone wants to know. But they can stay home.

As for me, while I plan my next journey out, I try to remember Monomoy and face the really scary business of day-to-day living with purpose and a sense of my own necessity, as the birds and animals do.

With me I carry a page torn out of *Crossing Antarctica,* the journal of Will Steger, the leader of a six-man international team that crossed the vast southern continent on skis and dogsleds—and faced dangers more tangible and extreme than I probably will ever know. In the long polar night, in the midst of his expedition of hardships, he recalled the earlier difficult and rewarding times: "During the struggle to raise money to go to the North Pole," he writes, "we had an ardent supporter in Duluth, Minn., an 85-year-old woman named Julia Marshall, whose family owned a hardware store. At a time when we were desperate for cash, I remember getting a check in the mail from her for $5,000. Accompanying the check was a nearly illegible note, which took me four or five readings to decipher. It said simply: 'WE NEED ADVENTURE NOW.'"

And we can have it.

Of course, adventure, like everything else worth having, has its price: I've had the discomfort of poison ivy for weeks; I know what it means to be cold, drenched to the skin, and squirrelly from cabin fever. But a little risk has its undeniable payoffs, too: being awakened at midnight by the eerie, lone cry of a great horned owl; being stopped dead in one's tracks by a doe diving through bayberry for cover; finding all vital hungers filled.

Talk about fear. You could move without love, forget how it feels to live. You could think you were safe—and never know the danger of deep joy, the pitfalls of beauty, the passion of being free.

The Stride of Sand

We may never unravel the interworkings of the long slow rise of sea level and the daily work of wind, wave and tide. Yet gradually we are learning how the beaches survive, so that the story of a particular event, limited to a short stretch of coast and span of time, can be told.

—WALLACE KAUFMAN AND
ORRIN H. PILKEY, JR., *The Beaches Are Moving*

In the middle of winter, it took seven white-tailed deer and a snowy owl to revive the beating of my wild heart.

They were the small blessings that came with the ritual return to Monomoy.

I had sojourned at the keeper's cottage twice, had seen a second winter come and almost go, when one day in late February—earlier than ever before—Hillary and I attempted a trip. Midwinter is still a cold, harsh season on the outer beaches, but we were restless and decided to brave the conditions anyway. We dug out long underwear, fleece vests, down jackets, and rain gear and set out at dawn on two unseasonably warm mornings in a week.

It could have been May the first day we made it to Monomoy, the weather was so placid and bright. But the odd silence didn't fit; the migrating birds weren't back. Gulls, geese, and ducks still dotted the beaches, and near the marshes west of the lighthouse, crows were on patrol, taking turns harassing the raptors, mobbing the northern harriers, and, in one instance, a great horned owl.

I spent most of the morning walking alone, crossing the washover at the island's midpoint and wandering along the ocean side of Monomoy all the way to Hospital Pond. The seals were sunning and sleepy, as I expect I would have been if I had stopped to lie down on the shore and to be lulled by the surf. But I wanted to make the northern end, to see how the island was holding up in the winter sea. Nothing was familiar; the sand had moved on, as I had, pulling history and topography in its wake.

Clouds and a bitter chill cast the tone the second day that we came, and having landed, I set out in the opposite direction, heading toward the lighthouse, hiking in over the sandy trail from the Sound through the marshes. By most accounts, the damp, raw day would have been less than ideal for being on Monomoy. But, in fact, the harsher weather made everything about the day seem more fierce and difficult and right—the ragged edge we tend to associate with wilderness.

I had barely broken onto the trail when I saw the deer. They were so far off, they registered only as blurred forms floating through the brush and high grass. Even with binoculars, the best I could do was watch their backs as they disappeared down the island, their white tails bobbing like signal flags above the foliage. I kept to the path, checking for known landmarks—a deer skeleton at one turn in the trail, recognizable dips and mounds in the sandscape near the Sound—and listened to the unrelenting wind all the way to the steps of the light. I rolled my pack off my shoulders onto the deck and threw myself down, too, leaning my back against the gear and putting my face full into the gusts. I watched the marsh for a few minutes, then went to check the few trees for nests.

I had half-expected to find evidence of a great horned owl—one has been known to nest near the lighthouse, and in the past I had found pellets, feathers, and the bones of prey to mark the spot. I had even seen the owl itself holding court from the pines.

But this time, it was the snowy that caught me off guard. I was nosing around the trees, my face to the ground, looking for clues, when I heard a rustling a few feet away. I lifted my head, and the great white bird rose, as silent and steady as my gaze, almost into my face before turning into the wind.

The tides were turning, too, by then, and soon, I knew, Hillary would be casting about, trying to find me, so I started back out to the Sound. Halfway there, the deer crossed my trail less than twenty yards off—seven of them stopping and staring at me as I stood in the path. They watched me for a minute or more, and then, as quickly as they had come, they dissolved into the moors.

At the shore, Hillary was still scratching for quahogs. As I dawdled back down the beach, desultory as wrack, I noticed he had picked up another tag-along straggler. About thirty feet away from him, a harbor seal breached and dipped in the waters, intent on the man—who, nearly chest deep in water and dragging a rake across the bottom, appeared, in his own way, to be bobbing in the surf. Hillary, at first oblivious and too hard at his labors to notice the mammal, finally glanced up and met the seal's gaze—but only for a split second, before the creature flipped over and dove beneath the surface. Within seconds, it was back up, no farther away, just occupying a different point on the circle of attention cast round the fisherman. For a half hour the seal kept up the curious pirouette, as if anchored there, floating around the man as though the two of them were bound by some invisible net.

Then, without warning, it disappeared, and we quit for the day, pulled by other ties back to a near shore. The eider fled before us, and the Sound gathered our wake into itself, the muted hard light of midwinter awaiting us on land that lay, it seemed, in a wholly different world than the safe sanctuary of wilderness.

But the barrier island followed me through short days and long nights, sifting into my heart, just beneath the level of consciousness. One night, about a month later, when the weather in the outer world would have made actual passage to Monomoy impossible, I dreamed I was walking the outer circle of the island. Outside, the dead of winter locked the mainland in quiet, grass and reeds brittle as parchment, bayberry thickets and sedge like skeletons rustling in the bogs and marshes. But indoors, an hour before dawn, sleep carried me to Monomoy.

In dream, I retraced my regular passage north along a stretch of the ocean side of South Monomoy. I passed a recognizable berm, the beginnings of dunes, large and small ocean debris, driftwood, and peat mats. In the phantasmagorical walk, on the island mapped by haphazard impression, I moved along, idly noting again things I had seen in the past, until I arrived near Hospital Pond. I glanced ahead to see how far away the marshes near Inward Point might be and saw at the edge of a low dune something unexpected: the jawbone of a whale.

I moved on until I reached the huge bone, unmistakable by size alone, and exposed in the sand. I tried to lift it or move it, but it was immense, too heavy and cumbersome to manipulate. I threw all my weight against it, trying to roll it, but I could not pry it loose. It barely rocked, absorbing the force of my insignificant determination and strength. I spent what energy I had, then gave up, and sat down on the gigantic remnant that had been distressed to the color of ivory by the polishing of moving sand, sun, and water.

I awoke, the sense of the island, the sea, and the sky still vivid and haunting; and though it was not yet daylight, I felt compelled to get up and record it. Just before dawn, I climbed into bed again and fell back to sleep.

Winter wore on and the sharp edges of the dream eroded. But very early in the spring, I had a chance once again to get out to Monomoy for a morning, and I made a sweeping arc over the island, stopping to check on conditions at the lighthouse and then swinging over the dune lines to the ocean beach. I intended to circle north over the tip of the island to the Sound and back to the boat, but as I approached the growing expanse of marsh and tidal beach between North and South Monomoy, a chip of white stood out in relief against the sand.

I picked up my pace and made my way closer, until at last I was standing over it—the fifteen-foot-long jawbone of a whale. I turned myself over to the vision from weeks before, reenacting the futile attempt to move the bone, digging along its smoothed edges until it sat in a cleared furrow of beach. But I could not lift it, could not dislodge it from its casket of sand. At last, I gave up, and instead hovered for a while, using the jaw for a perch and combing the near sandscape for signs of any other disgorged bone. None was revealed. I made mental note of my location, then left, retracing my route.

Spring warmth came early that year, and a few weeks later, I was able to make another crossing to Monomoy. But the bone was gone, reburied, I assumed, above the berm. Certainly no one else—other than the conspiring ocean, sand, and wind—would have been any more prepared to transport it, especially along the ocean shore and shoals. Still, the disappearance of a bone weighing hundreds of pounds was as startling as its discovery. Now you see it, now you don't.

I do not make much of my premonition about the bone, except as evidence of an uncontrollable bond with the island. Already the sands of my life were drifting in inexplicable ways, defined by the tides and dune cliffs of Monomoy.

But chances are that no other place could have produced the conditions of the dream more realistically than a barrier beach. The transformation of the shore and the excavation, then reburial, of the bone in such a short time reinforced what I already knew of the elemental laws of life on a barrier island. Everything moves, always changing and being changed. And it is churning sand, the form and substance of Monomoy, that expresses the principle and the place.

IN THE MOST RECENT MILLISECOND of geologic time, Monomoy has undergone constant change as a landmass—from a series of small barrier-spit islands in the 1800s to, in this century, a peninsula connected to the mainland town of Chatham, then a single island, and most recently, to an island split in two. The entire "barrier island formation," as geologists call the type of configuration of sand that characterizes

Monomoy, rests on a bed of glacial material that was left in the wake of retreating glaciers an estimated 18,000 years ago. It took another 12,000 to 14,000 years to form Monomoy.

Atop the glacial deposits lies the restless sand that is Monomoy, supplied by the erosion of the ocean side of Cape Cod. As points farther north along the Outer Cape are eroded and washed south, the Monomoy islands accumulate sand. Depending on episodic patterns of erosion and accretion, Monomoy at different times will become slightly larger or smaller. But the exact size and shape of the island are never still, never constant. Not only do daily tides effectively extend the reach of the tidal flats of North Monomoy and the ocean beaches of South Monomoy; the action of waves refracting around the tip of South Monomoy has, over time, caused its shape to modify dramatically and relatively quickly, even in terms of the human calculation of time.

Waves move the sand southward around the end of the island in a J-shaped pattern, until at last the hook closes in on itself, and the process repeats. The numerous brackish and freshwater ponds at the tip of South Monomoy reflect this continuous reshaping and sculpting process, but no more so than the position of Monomoy Light. Once located near an earlier dune line at the southeastern end of the island, the now-defunct lighthouse, because of the movement of sand and pebbles around it, sits a half mile inland from the ocean to the east, and Nantucket Sound to the west, and more than a mile from the point. The tip of the island literally has retreated from the light.

During the last 350 years, the island has migrated steadily south and west, eroding, altering itself, growing in places, then slipping away. Maps from Monomoy's past prove the point. The explorer Samuel de

Champlain's depiction of the elbow of Cape Cod in the early 1600s includes a long, narrow peninsula, Port Fortune, running south from the mainland for only a short distance before turning sharply and extending parallel to the Cape's southern side. A little more than a century later, a mariner's chart shows a single, narrow sand mass, lying in the present-day location of Monomoy; it has split away from "Nauset Bar." Wreack (Wreck) Point, at the cove in which an early tavern had been established, also appears on this eighteenth-century rendering, but no human structures are noted. Included with the identification of the island is the handwritten notation "fine brown sand," a designation to guide those navigating around the island to hospitable landing sites.

Maps based on records from the mid-1800s show the Monomoy of 150 years ago cut off from the mainland at Chatham. But the fishing community, Whitewash Village, is marked at the Powder Hole harbor near the southern end of what would later become South Monomoy, and Monomoy Light is shown on the oceanside beach at the southern tip. Maps from 1868—merely fifteen years later—reflect the erosion and accretion of sand and its impact: A continuous sand spit runs from north of Chatham to the tip of South Monomoy, and the Powder Hole harbor is disappearing under the swirl of sand around the island's southern tip. The island itself is still severed from the mainland.

Shortly after the turn of the century, however, that rift is closed, and by 1917, Monomoy has become a peninsula jutting south from Stage Harbor in Chatham. Already the build-up of sand at the southern end of the island has effectively changed the location of Monomoy Light. The sand has caused the shoreline to migrate, and the light is no longer at the edge of the island, but rather sits behind a line of dunes. The hills of sand will continue to pile up, creating secondary dune lines further segregating the light from the shore and, beyond, from the shoals.

Fifty years later, in late March and early April 1958, a severe nor'easter breaks the landmass from the mainland once again, this time separating the Monomoy refuge from Morris Island, in Chatham. Two decades after the first breach, a brutal blizzard hitting the coast in early February 1978 rends the island wilderness in two at the vulnerable, narrow point of Hammond's Bend, near the contemporary northern edge of South Monomoy. Left behind are two islands with a tidal inlet between them.

Since then, sand swirling around the southern beach has transformed the Powder Hole into a large, shallow tidal pond and marsh and has expanded the southern end of the island, now a bit more than a mile wide and the broadest stretch of landscape. Cartographers, monitoring the movement of sand over centuries, have left records indicating that the northern third of the island today has been repositioned more than a mile and a half from its location 350 years ago. And geologists, looking ahead, predict that within twenty-five more years, South Monomoy will break again at its approximate midpoint, or North Monomoy will reconnect with the mainland at Morris Island, Chatham, or with the southerly encroaching Nauset Beach. One thing is certain: Whatever is there now won't be what is there later—in a week, a year, or a century.

Now you see it, now you don't.

SINCE YOU CAN DISCERN no sure beginning and can deduce no absolute end, you start, arbitrarily, to guess at your location by casting about where you find yourself—in the middle of things. First, you try to set the contours of the near terrain and project the farthest boundaries of the place. Failing even that, you search for whatever seems to last.

On Monomoy, what endures is grand restlessness and change. Even if you don't know it at first—and I didn't in the beginning—you start to get a sense of the place as dynamic, creation constantly re-creating itself. The ocean heaves against the long shore, the dunes shudder, and the coastline begins to shatter in tiny pieces. The reach of tides and the spheres of heaven set the boundaries of night, day, and, again, the dark. Time and space belong to the birds, which hover and dip, coasting on currents of histories we do not see. Human marks pass away, consumed by the sprawling sand, and what little is left—the scant trail threading through the Hudsonia moors or a tower erected in the face of the wind—is mute witness to our reluctance to leave paradise to the grasses and birds, the snakes and the scrub.

Of all the stirring landscapes in which to find yourself, a barrier island can be one of the more disorienting because there are no truly fixed points to count on, only patterns to predict the unrelenting movement of sand. And we, too, move as the sand does, swept along by greater storms, of sea and savage civilization. Like the waves and tides, we come and go, mounting the shores and retreating from the face of the wind. Yet the vicissitudes of life on Monomoy are constancy enough for the creatures and the vegetation eking out a life in the sand, sun, tides, and rain. During spring, summer, and fall, the two islands host a herd of deer, small rodents, snakes, toads, and insects, as well as scores of species of birds. Gulls, often feeding on the mainland, taking what scraps and garbage they find by day, make the return flight to Monomoy by nightfall and dominate the skies over the refuge. But many other birds are commonly sighted on this major coastal stop-off point for migration, including snowy egrets, herons, common terns, oystercatchers, teal, cormorants, mergansers, and black ducks.

As a wildlife refuge, Monomoy exists for wildlife first and for humans only incidentally. The islands' designation as wilderness means "management" should be limited to minimal intervention into the cycles of natural life, even when it might seem beneficial to wildlife. There have been striking exceptions to that rule, of course: In the 1950s, the Fish and Wildlife Service eliminated from the refuge many of the small mammals that preyed on birds, including foxes, raccoons, and skunks. Freshwater marshes and ponds were bulldozed into areas of South Monomoy to enhance it as a habitat for waterfowl. Gull control has long been a part of the overall management plan, but in more than a half century of wilderness management, the Fish and Wildlife Service has put the most energy into curbing human disturbance and its damaging effects on the birds and animals. Traditionally, officials have tried to emphasize public education about the islands' natural history and resources over law enforcement.

As a result, parts of Monomoy are visited by people all year long, but mostly the islands have returned to a largely untouched natural state, an uncommon wilderness within reach of Cape Cod, one of New England's most popular—and, in summer, congested—sites for migrating tourists.

Monomoy is a place where novices and world authorities in ornithology can find breathtaking birding, as well as a microcosm of diverse habitats to study. And because the islands change from day to day and week to week, there is a novelty about the place that never wanes. You can ground yourself in what you think you know—how and when the seasons pass, the hour of the day dawning or the night falling, but the heart of Monomoy seems to beat with a pulse all its own.

In a log of lighthouse visits kept at the keeper's cottage by staff of the Cape Cod Museum of Natural History, the list of identified species on any given overnight trip stands as the most straightforward record of the island's significance to birders, botanists, and wildlife biologists. Pick any dates—for example, October 8 and 9, 1989, when a naturalist recorded the sightings of "5 white-tailed deer, 2 shrew, 1 meadow vole, 1 white-footed mouse, thousands of yellow-rumped warblers, 50–60,000 tree swallows, 25 black-crowned night herons, 3 to 5 great blue herons, 60 ruddy ducks, 20 scaup, 75 green-winged teal, 200 blacks [black ducks], 30 mallards, 6 gadwall, 12 wigeon, 43 snow geese, 2 peregrine falcon chasing teal over Big Station Pond, 1 merlin, 2 common loon, 3 gannets, 1 sapsucker, 1 white-winged scoter, 10 shovelers, 40 blue-winged teal, 15 pintail, 40 Canada geese, 3 palm warblers, 1 fox sparrow, 12 harriers."

Such sightings are unequaled in virtually all other parts of the Northeast and are among the everyday natural phenomena that create Monomoy's mystique. Even for the people whose job it is to protect the refuge from unwitting—or willful—misuse by humans, the wildlife sanctuary remains a place of almost primal beauty and drama. On Monomoy, people often say, you feel as though you're standing at the dawn of time, watching the creation of the world.

I DIDN'T HEAD TO THE ISLANDS with the intention of arriving at a place unsullied by human presence, but looking back, I see that Monomoy gave me something I had desired so deeply that the longing had gone beyond words, to the unspoken core of my being. It was a yearning for innocence, for a present defined by something other than past ruin, for a future that might open onto hope. But I wasn't thinking about any of that when I took to Monomoy. I simply knew that something about the place, its sounds comprising water, wind, and an uproar of birds, brought me peace.

Often, during my time at the light, mornings dawned humid and foggy, and the day worked up to glaring heat. I would awaken to a landscape heavy with dew, at that particular hour that seems to be its own season—a cool, wet, perfect moment between summer and fall. The dew would be clinging, still, to the tall grasses and sedges, and the world belonged to the fork-tailed barn swallows in the air and the orb-weaving spiders with their nets glistening in the fog. It was weather that could make me weep, but for what I did not know.

From time to time, I would try to discern what it was about the moist, heavy thicket of air that could start a shudder moving deep inside me, as though I were being transported into meditative ecstasy or solitary bliss. Sometimes, it seemed as though the feeling was a fragment of memory, a long-forgotten recollection awakened in senses stimulated beyond reason. Something in the atmosphere was familiar, an unbroken relation as close as a blood bond. But it might have been that the sensation was just about water. It could be explanation enough to remember that I have always lived on a great body of water—one of the Great Lakes or along the continental edge, where everything important yields, in time, to the sea. There probably are some weird, uncharted connections in these spots between meteorology and mood swings, dispositional rises and falls that parallel lake-effect weather patterns or oceanic swells.

But I am guessing that the explanation runs deeper, to the chemical origin of things, the primal, biological starting point. Water is the compound that links me to my home in the cosmos, the Earth, the water planet. The substance that constitutes so much of the physiological "I" covers almost three-quarters of the planet and is the still-secret storehouse for some of its greatest mysteries.

They say that God is in the details, and redemption and release are there, too, even though I am seldom looking when they strike me most. On mornings like those on Monomoy, water, like air—another life-sustaining element I take for granted most of the time—impressed itself on me in small, cogent impressions: droplets reflecting and refracting the early lights, mist setting spiders' silks in muted relief, curtains of fog drifting through the marsh. The water, softening the usual haste of morning, could wash away a false start, it seemed, and redirect my energies along their proper, appointed course. Standing in the wet grasses at daybreak, enveloped by vapor, I would feel an intimation of my true center. The power of the spirit, the essence of life, has been cast in metaphor as air and breath, its authority as pillars and tongues of fire. But in those moments alone, in the clearing fog and heavy dew at the light, I sensed something else, something different—the flow of eternity, the fluid truth of being, and belonging.

If you were to go looking for the point at which Monomoy begins, you would likely start at the most northern edge of the refuge or on Morris Island, in Chatham, on mainland Cape Cod. From either vantage point, you could see how sand is carried down the arm of the Outer Cape to its elbow, and beyond. But it is the changeable center of South Monomoy that provides perhaps the most accessible setting of the island refuge as a world. At roughly the midpoint of the southern island, the land tapers to less than a quarter mile, and it is slipping away daily. There, at high tide, beachcombers making their way from the southern tip's ponds and inlets to the marsh at the northern end can observe the forces that set the boundaries of the island at any given moment: Nantucket Sound the inner limit, the Atlantic the outer. At key points of high dune elevation or eroded, overwashed beach, walkers can look to their right and glimpse the surf foaming on the outer beach even as short, shallow crests reiterate themselves on the tidal flats of the sound to their left. And at the midpoint, as perhaps nowhere else on the island, it is possible to regard—in a sort of natural cross section of the place—the dynamic equilibrium of energies in wind, waves, and tides that define Monomoy and the stark, sovereign elements that limit it: sun, sand, and sea.

Dynamic equilibrium—a balance of several natural, changeable forces—is the closest thing to stability or constancy on a barrier beach. It involves the interplay, or exchange, of shaping influences among materials (sand, silt, stone, biological or organic debris, and refuse from the sea); the energy of wind, waves, and tides; the contours of the beach at a given time (its breadth and slope); and sea level, that is, whether the landmass is rising or falling and whether the ocean is gaining or losing water. It sounds straightforward enough, yet it is anything but simple. Part of the mystery of dynamic equilibrium that governs barrier-beach topography is that one aspect—say, the long, slow rise or fall of sea level—might seem insignificant over the course of a human life or the experience of a generation, while another—the daily action of wind, waves, and tides, especially during storms—might appear dramatic and pronounced to someone observing a shoreline even for just a season. Add the complication of a toppled mast, the beam of a shack felled years before, the sprawling skeleton of a humpback, or the crawling intrusion of human development, and the picture is altered again, this time by obstacles that reorganize the shape of drifting sand. The way the process of dynamic equilibrium plays out on a barrier beach and its associated ecosystems of tidal marsh and dune is intricate and complicated; and geologists are still probing the secrets of how beaches are built up and broken down, sculpted and reshaped, eroded in one place and resurrected down the line.

We do know some things, though, or at least we accept certain notions about how barrier beaches are formed and changed. The idea of "spit accretion" is the most widely accepted interpretation of how Monomoy and several other New England barrier beaches, including the vast Nauset Beach to the north, were formed and continue to exist. According to this theory, sand is generated by the erosion of headlands, or offshore glacial deposits. The sand sweeps along the coast and over time forms a long finger or narrow arm of "land" connected to the mainland but jutting out into an ocean or bay. Behind this sand ridge a lagoon develops, and possibly a sizable salt marsh, too. The process continues, never static, vacillating according to seasonal and meteorological conditions—such as variation in wave action and sand deposits from summer to winter or the tempestuous changes caused by nor'easters and hurricanes. Meanwhile, the sandscape literally migrates, pressed onward by the drifting sand and other material "stored" offshore. The terrain, forever changing, forever young, persists, and over the long haul of natural time, becomes stable enough at least to support habitats for a variety of plants and animals. Ultimately, dunes will form above the beach, and thickets of salt-hardy trees or moors of rugged grass and other plants will take hold.

The evolution and resilience of barrier-island habitats is especially evident on Monomoy. The island wilderness is at once substantial and small enough so that a visitor can read the record from atop a high dune, or better still, from the peak of the lighthouse. From that lookout, it is clear that the southern and northern reaches of South Monomoy host the more intricate and diverse habitats that a greater expanse of sand (temporary shelter from the sea) and vegetation (an illusion of permanence, despite the erosive winds and water) permit. At the extreme reaches of the island, freshwater and brackish ponds and saltwater marsh support forms of life that could not thrive given only the scant sustenance available at the narrow midpoint. Yet, halfway between Monomoy Point, which overlooks the confluence of ocean and Sound, and Inward Point, some three to four miles north, the more commonplace—but still ingenious—aspects of the island's ecosystems show themselves.

The sand itself is a study, a canvas upon which the ocean, storms, high winds, tides, and surf perform an ever-advancing pentimento of current design layered over past artistry. On a day in early September, shortly after noon, a bracing gust blows from the south, whipping the beach grass and pressing the blades and tips to mark out compasslike wind rings around their stems. The swash—or rush of water from breakers—and, at some points, the overwash—sea water spilled earlier across the width of the island during storms—leave their own imprints, mingled here and there with drift lines and debris. And bisecting the length of the narrow island midzone are marks not of water or wind: These precise clefts come from one of the deer in the island's small herd, left as the animal moved across the sand from the southern grassy moors to those farther north, nearer the marsh at Hospital Pond.

But all these signatures are impermanent at best, subject to change instantly or erasure overnight by winds and water, whether the force be driving surf or regular tidal movement. Only the more durable central corridor of this strip—which approximates one of the island's past dune lines—displays signs of flora that require at least slight stability to last: Here dusty miller, also known as "beach wormwood," has established itself along with seaside goldenrod.

These two tough plants often spring up in the same zones of barrier lands because they have evolved to adapt to the brutal, desertlike conditions. Dusty miller, for example, was originally an ornamental species, introduced to the New World for growing in poor soils and rocky, arid land. A plant that appears silvery against the backdrop of the sand, it is covered with fine white hairs that both prevent moisture loss and serve as reflectors to the sun's rays—thereby enhancing its chances of survival. Similarly, seaside goldenrod is salt resistant and grows from a clumped root system that sends forth several stems simultaneously. Each might grow to five or six feet tall, all the while bearing the full assault of sea spray and, in some places along the island, overwashing waves.

The two plants are key vegetation for developing dunes, each in its way capable of serving as a sort of shield against which blowing grains of sand can be captured and held in place—the earliest suggestion of contours and terrain that eventually may build up. In fact, seaside goldenrod and dusty miller frequently are found growing together, their roots snarled into an exponentially stronger, shared system, capable of securing not just the plants but the habitat itself.

But most plants and many animals cannot tolerate the merciless conditions and the uncertainty of the midisland environment. The noticeable lack of vegetation and the apparent absence of diversity of life demonstrate that everywhere, anywhere, is a shore on which life is always a struggle.

Even "shore," however, is a negotiable boundary, altered twice each day by the tides and erratically by winds and storms. The tidal flow—the rising and falling of the oceans' surface and that of connected bodies of water—is governed by the gravitational pull of the moon, and, to a lesser degree, of the sun, on the revolving Earth. And Monomoy's shores and surrounding waters are drawn not only by tidal patterns of the Atlantic but also by tidal variations in Nantucket Sound. Tides are the source and sustenance of life for many organisms, alternately providing water and air, which translate, on the one hand, into protection from drying winds, sun, and heat and, on the other, into a constantly freshened, replenished environment in which to live, reproduce, and daily resume the search for food.

Natural scientists, trying to distinguish between formations of land and sea and the forms of life that inhabit them, have taken their cue from the tides and designate different areas, or zones, whose demarcations can be traced to tidal flow. For example, the area that is exposed between high and low tides is known as the intertidal, or littoral, zone; while the stretch of shoreline above the high tide line is termed the splash, or spray, zone. Below the low-tide line begins the subtidal zone. Each has its plant and animal residents, whose life cycles mesh with their environment in a complex interdependency so balanced and precise it humbles human intellect and invention.

Take, for example, something ordinary: the grass. For miles on end, down the island's spine, the dominant vegetation is beach grass. This hardy grass, which goes from a glistening green in summer to a muted yellow in fall, is uniquely suited to Monomoy's harsh conditions. For one thing, it requires little nourishment—only the nutrients stockpiled in new sand—and it exploits the wind-blown environment even while helping to stabilize it. In fact, moving sand triggers the spread of beach grass, and if the beach or dune around it ceases to move, it stops growing. Then, as the beach or dune resumes its drift, the shoots of the grass and its roots once again stir to life.

The shoots begin anew to capture grains at their base, forming tiny mounds, which foreshadow the dune to come. The shape of the stem and blades both reduces the speed of wind carrying sand close to the ground and catches the grains collecting near the base. Eventually, the stem is buried and simultaneously is transformed into the new growth medium for the underground stem and root system. The scant nourishment the grass needs and can derive from blowing sand is enhanced by its own compost, the organic matter it contributes as it is buried.

At the same time, the specialized creeping root and stem system of the existing beach grass is stimulated. Running both vertically and laterally, the roots enable the grass to continue growing despite the encroaching sand, which may build up to as much as three or four feet per year. In this way, the expansion of dune and grass is inextricably linked; but if the dune ceases its spread, the beach grass will languish, finally lying down, and over time, decomposing—opening a potential niche, supplied with organic material, for other plants and shrubs.

Unlike beach grass, which exists in an interdependent relationship with its environment by taking advantage of the severest qualities of life there, many animals of the shoreline endure because they have developed techniques to avoid the rugged elements. Much of the animal life of Monomoy is, for example, nocturnal, shunning the almost desertlike conditions of day for the cover of night. At dawn or near dusk, Fowler's toads, garter snakes, and a variety of species of mice and voles are visible; but the activity of all these creatures accelerates at night.

This avoidance of hostile elements is characteristic of life in the intertidal zone, where burrowing is employed by tens of thousands of creatures every day, as they elude the lethal drying of sun and wind and the tumult of the waves. Here, the riot of life is under the wet sand: quahogs, steamer clams, razor clams, and many varieties of sea worms—to name but a few organisms—enact their life cycles. They have evolved to flourish here. The soft-shell clam and other bivalve mollusks possess flattened shells that slice like sharpened blades through sand and under water, easing locomotion. Likewise, they have long necks, or siphons, that enable them to remain hidden and protected under the sand while scouring water for food and oxygen. And they are equipped with a large gill chamber inside their shells that facilitates respiration even in the dense, heavy underworld of soaked sand or mud.

If you were to walk, at low tide, from the tidal flats on the Sound side of South Monomoy across to the Atlantic edge, you would pass through a series of areas that, on first glance, might seem all of a piece and hardly distinct. But closer observation would reveal a compact microcosm of thousands of years of habitat formation and species adaptation. Time is eerily telescoped—it is as if you might be able to walk over the sand, back through eons, and glimpse the islands' origins, then continue, marching through identifiable habitats, back to the present on the other side.

Bay (or in this case, Nantucket Sound), tidal flat, bay shore, and salt marsh form one broad series of zones. Governed by the ebb and flow of the tides, this area grows and diminishes twice daily as the water rises and falls. At its greatest expanse, which occurs at dead low tide, this area is characterized by sand flats, mudflats, and marsh. On South

Monomoy, this environment includes the marsh at Hospital Pond and the tidal flats at the northern end of the island, as well as the tidal shallows and pond of the Powder Hole to the south. In this tidal strip, you encounter herons, egrets, gulls, sanderlings, sandpipers, terns, and a variety of other shorebirds working the eel grass, black wrack, flats, and shoreline, scouring the shallows for food. On and in the wet, rippled sand you discover a variety of crabs, numerous mollusks, and sea worms.

Dune communities occur as you move inland, passing through back dune, secondary dune, trough, and primary dune habitats formed by the accretion of shore material washed down from the north and shaped by winds and waves. Walking the dunes, you cross over the highest elevations—all erected by the combined labor of sea, wind, and sand. Here you can observe the low grassy growth as it surrenders to more dense shrubbery, higher dune grass, annual and perennial plants. These ecosystems support more diverse animals, such as deer, muskrats, snakes, toads, spiders, grasshoppers, and owls.

On some barrier islands, the landmass becomes increasingly stable over time and will allow for the growth of a maritime forest. Only the broad southern end of South Monomoy, with its large ponds, has persisted—and grown—to the degree that forest could be sustained. Even so, the closest thing to woods on Monomoy are thickets of shrubs, primarily bayberry.

The beach—comprising back shore, open beach, and foreshore—is the point at which the variety of flora and fauna changes dramatically, paring down once again to cast-ashore seaweed and shells tossed and ground by pebbles, waves, and sand. Some shorebirds frequent this habitat and may move back and forth from tidal flat to ocean shore to feed. Evident here are such organisms as beach fleas, sand flies, and a variety of spiders.

The ocean is its own sphere. Spread out beyond view is Georges Bank, the underwater edge of the continent. It is, effectively, a still-alien world, that of the sea. Marine animals, including seals, dolphins, and whales, offer the mammalian links to the waters from which we crawled, in our earliest incarnations as a land species, millions of years ago. But it is the authority of their ecosystem—the vast ocean flowing over the continental crust off the East Coast—that determines the fate of the nearly three hundred islands hugging the coastline from Maine to Texas. With its 2,700 acres, Monomoy occupies only seven and a half miles of the 1.6 million acres and nearly 2,700 miles of barrier islands buffering eighteen states. In contrast to this tiny island refuge, which has been resurrecting itself for 6,000 years, the space of one human life—all the experiences, memories, and time—is a small occupation indeed.

Break 1978

Don't fear your hot tears
cry away the storm, then listen, listen
—Pat Mora, *Chants*

In what turns out, on reflection, to be one of those embossed memories, with all the details of an instant in childhood lifted in sharp relief, I am where I have been left, standing in the corner of my parents' bedroom, alone, naked, maybe eight years old. I am waiting—on what could have been any Sunday of many, except for a few riveting particulars—waiting a long time, it seems, for someone to come for me, to find me there, alone and with nothing—nothing on, nothing doing.

From a long way off, from downstairs, I hear the muffled voices of my brothers and sisters, laughing and bickering, as they set the table for dinner. *At dinner, then, I think. At dinner they will notice I am gone and someone will come for me.*

I hold my breath and listen, the disembodied voices drifting through the hot air registers, up the carpeted stairway doubling back on itself, the way the mind does, climbing through the thin air of sensibility, choked by the overcome senses, the weight of too much, too much. I hold on to what little I can, hugging myself in the cold room, rubbing my thin arms with my hands, up and down, to feel again something like warmth. Far away, I hear the chink of silverware, and the plates, the squeal of glaze against glaze, the thud of their fall as they are tossed on the tabletop, the pad under the cloth deadening the sound.

Before long I am altogether gone in the things of the room, garlic and cinnamon climbing all the way from the kitchen to here; the garbled, broken words floating up from family, from below; the bedroom slipping away, faint, unsafe—*come back, says a soft voice inside, stay here*—the clammy wood of the floorboards under my feet, the brittle chill of the wallpaper against my back, my toes curled tight under themselves, nothing beneath me, nothing all around.

I will remember this, though I find nothing here. No one here. Not even myself. Only the debris of memory. Feeling sealed in the shattered fragments of things, glimpses of fact I can take as I soar and fall, at once rising and dropping off, away, out of myself, out of this place, as subdued as voices and easy laughter from somewhere else. As though nothing at all had happened. To no one.

At the hard heart of the continent, I wait. For nearly twenty more years I wait, before silence breaks.

THE TURMOIL FAR OFF went on for days and days, the swirling snow, gusting winds, and surging seas punishing the northeastern seacoast. Twenty-seven inches of snow fell, human activity staggered and slowed, leaders declared a state of emergency that would prevail for five full days. Regardless, the wind and waters had their way, claiming sixteen lives in Massachusetts, forty-five in New England.

Storm tides clawed at the Cape, swelling to fifteen feet at the Provincetown tip and flooding the streets of the commercial district there. Two days into the blizzard, the half-sunken tanker *Pendleton,* embedded in sand a mile east of Monomoy, shifted its position. It strained under the haul of the seas for two more days, until finally it disappeared altogether. The island itself was ripped in two at Inward Point. Where earlier overwashes had threatened to run the sand quick and thread a breach of several feet between the northern and southern portions of the spit, the ocean's fury now wholly sundered the small isle. When the waves settled and the wind went calm, two islands remained, separated by a half mile of sea.

At the center of the continent, I navigated the edge of the Great Lakes, preparing for the emigration east, first across a state, and later across the continent. I left my little signature and moved on, not knowing where the wandering would take me, or when. Uprooted by my own eclipsed history, I began the exodus that would take two decades to complete.

The Human Countenance

I think of the people who came before me and how they knew the placement of the stars in the sky, watched the moving sun long and hard enough to witness how a certain angle of light touched a stone only once a year. Without written records, they knew the gods of every night, the small, fine details of the world around them and of immensity above them.

—LINDA HOGAN, "WALKING"

THE FIRST HUMAN INHABITANTS on Monomoy migrated from the far north and west, and later from the south, prompted by changes that the climate wrought on the land. They settled where food and shelter were theirs for the taking and, in time, moved on again.

Those who stayed shared their name with the place and the light—*Wampanoag*, meaning "people of the dawn," "place of the dawn," or "people of the white wampum."

Like most American Indians, the native people of Monomoy, it is thought, descended from Asians who migrated between twelve thousand and thirty thousand years ago, possibly in two distinct waves, across the ice or land bridge that joined Siberia and present-day Alaska. As the last great glaciers retreated and weather became warmer and more tolerable to the south, these early native peoples, or Paleo-Indians, kept moving, eventually making their way to western North America and Central America; and their descendants over the next several millennia pressed on farther still, finally reaching the Northeast and the Atlantic to the east, and Tierra del Fuego to the south, at the far edge of Cape Horn. Those who ultimately reached the shores of the Atlantic in southern New England are believed to have approached the region from the southwest. Humans traveled the path of forces greater than themselves; they trailed the animals who were pulled to new territory by withdrawing glaciers.

When tracing the early history of aboriginal peoples, what seems a stretch to the modern mind—namely, that people are inextricably linked to the land—is an inarguable starting point. Our contemporary speculations about prehistory flow from an understanding that the Earth itself is six billion years old and that *Homo sapiens*—or modern humans—arrived on the scene only about a hundred thousand years ago. History, as most of us think of it, covers merely the last forty thousand to fifty thousand years. By contrast, the earliest forms of life appeared in the oceans about 600 million years ago. Plants—which still are the foundation of life on Earth—had to evolve to live first in water and then on land. Only then did early terrestrial animals have places to hide and something to eat. In time, other meat-eating animals fed on the plant eaters; and evolution spiraled on, leading to early humans about four million years ago. But long before us, the dinosaurs, insects, reptiles, birds, hoofed herbivores, and ancient primates filled the land, the seas, and the air. And though we comprehend how linked other animals are to their environments, we have more difficulty seeing our own reliance on nature—especially in contemporary times in the developed world. We even understand that primitive peoples were dependent for survival on terrain and climate. The notion that humans could supersede, or dominate, nature could be entertained only after human beings felt some distance from its brutality and indifference. Survival preceded arrogance.

When the dinosaurs died out about 65 million years ago, the mammals rushed in to fill the open niche. Of course, in prehistory "rapid developments" could take millions of years to unfold. But life proliferated and diversified its experiments, and the continents moved and collided until by 3.4 million years ago, the land settled into patterns that we find recognizable. Early humans, too, look familiar.

But it would be a long time before existence would be predictable enough to accommodate the luxuries of art or philosophy. During the Pleistocene epoch, which began about 1.8 million years ago and continued until ten thousand years ago, weather still was erratic and changed drastically many times, moving and shaping land and seacoasts, which in turn governed what plants and animals could survive, and where. Ice was the biggest factor in limiting or extending human populations over the continents.

The Neanderthals are believed to have inhabited ice age–Europe and Asia from about 120,000 to 35,000 years ago, when they died out for reasons that remain obscure. Other *Homo sapiens* endured, and by 40,000 B.C., humans populated much of the planet, even building boats to carry themselves to the most isolated regions of the globe. In parts of Africa, Asia, and Europe, farming, boatbuilding, art, religion, and ritual had developed.

But in North America, such developments were delayed for another twenty thousand years. Until as late as eighteen thousand to sixteen thousand years ago, ice kept much of what is now New England locked away from human habitation. The continental shelf that corresponds to present-day Georges Bank—extending well beyond what was to become a string of barrier islands and spits along the East Coast—was still above water, and fossil evidence suggests that big game, including mammoths and mastodons, ranged along the coastline. Over the next few thousand years, however, the last glaciers pulled back to the north, and the sea level began once again to rise.

By about 7000 B.C., humans hunting caribou wandered over what is today southern New England. As the ice melted and the climate warmed, the landscape and everything in it was transformed. Tundra and grasslands yielded to shrubs and new grasses, as well as woodlands of spruce, birch, pine, and hemlock. And as the character of the land changed, caribou and other large animals moved east, following the grasslands and forest growing in the swath of the diminishing glaciers. Aboriginal hunters were close behind.

Radiocarbon dating of archaeological remains indicate that humans arrived in New England between nine thousand and ten thousand years ago, before Monomoy was even formed. The early natives of southern coastal New England appear to have been seminomadic, staying in any one area only as long as they encountered animals in the hunt or were able to gather other food. With the tempering of the climate, smaller animals and waterbirds arrived, while big game that sought out tundra and grasslands for habitat disappeared to cooler lands to the north. As the animals and the people who hunted them evolved, over a few thousand years, the social organization of human activity became more complex.

About ten thousand years ago, early peoples began to band together, and they wandered throughout the year, migrating to find the best conditions for encampments and hunting. Primitive tribes and their lands covered thousands of square miles—a measurement that would have been lost, of course, on them, since they located themselves *in* nature, not against or above it. When they described territory, they organized their thoughts concretely, not abstractly, and made distinctions according to natural landmarks—a pond, a hilltop, a grove of trees. "Property," and certainly "private property," was a concept not yet applied to land. They lived and moved, rather, within territories.

To survive, early inhabitants of New England and the coast could not rely simply, or even primarily, on hunting large animals for food and materials for creating clothing and shelter. They began to diversify their diet and with it their lifestyle, eating seeds, fruits, berries, roots, wild rice, rushes, and basic flours to supplement their gathered food. As they reached and began to live along the ocean shoreline, hunters turned to spearing seals and other marine mammals. On land, traps, pits, and weapons were developed for preying upon the last reindeer, and later, deer, as well as small game and freshwater and spawning fish.

Between three thousand and six thousand years later, glacial sediment, the rising sea level, and erosion of land farther north along the outer shore of what is now Cape Cod began to sculpt Monomoy and situate it as one of a long series of broken barrier islands running along the coast of what in several millennia would be the land between Maine and North Carolina. Human life and activity, like the land itself, entered a new phase lasting from 6000 to 4000 B.C. Human skills sharpened, tools became more sophisticated, techniques of hunting and fishing developed with more subtlety, intricacy, and risk as natives took to the open ocean in search of whales and other large marine mammals and fish, in addition to gathering shellfish and spawning alewives from the seashore and estuaries. Moving from place to place became less imperative and more dangerous and burdensome. Groups of people stayed more of each year at specific sites, and when they migrated, they returned over and over to places they had occupied before, cultivating communal living and travel.

Between 4000 and 1000 B.C., Monomoy was a distinct sand spit. Native people began gathering together in significant settlements during this period, and a variety of versatile tools proliferated. Weirs, for example, were long fencelike structures strung through streams or channels to snare fish and are still used today by fishermen along Cape Cod. Large bowls came into use, possibly for cooking, collecting, and temporarily storing food, but archaeologists do not know definitively what function these containers served.

During the next 2,500 years—from 1000 B.C. to about 1500 A.D., when Europeans reached the continent—native life in northeastern North America became increasingly more specialized, certainly in terms of social organization. Archaeologists have unearthed basket-shaped pottery bowls, suggesting more sophisticated food preparation, cooking, and storage than in earlier times. These bowls and containers were the first to be decorated, and designs indicate not only the luxury of aesthetics but also a sense of communal identity from one settlement, village, or region to another. Trade and some forms of currency—notably, wampum shells—began to surface, signaling that these peoples were aware of each other and interacted peaceably, as well as in warlike confrontation. Carved stone shapes of animals and other ornaments have been uncovered, as well as tools and weapons carved from antler and bone. By late in the period, farming was practiced, with squash, gourds, edible flowers, and, finally, corn as key crops.

The cultivation of corn, along with the invention and use of the bow and arrow for hunting, literally transformed the way of life for early peoples. By 1000 A.D., humans had become very skilled at killing animals—and other humans, if they deemed it necessary. Along the coast, many settlements still relied heavily on what could be gleaned from the shore and the sea—many kinds of fish, shellfish, waterfowl, and shorebirds. As lifestyles and activity became more diverse—in one place, natives focused on fishing, in another, on farming or at least on the cultivation of limited key crops—people again migrated to areas that were more suitable for the particular uses they found desirable. During the first millennium A.D., art, leisure, personal ornamentation, and intracommunal identity matured and deepened. Many artifacts point to use of tobacco, the production of distinctive clay pipes, more complicated food preparation, and even more attentive burial of the dead.

The Algonquians were the linguistic family of tribal nations that inhabited a region covering most of northeastern North America—Nova Scotia, New England, Long Island, and the Hudson and Delaware valleys. Their name is believed to have been taken from a Micmac (a tribal nation of Nova Scotia and eastern Maine) place name, *algumaking,* or "fishing-place," or a Maliseet (a neighboring tribe of the Micmac) word, *elakomkwik,* meaning "they are our relatives." But the Algonquians tended to refer to themselves with a collective name, *Ninnuock,* which translates into English as "the people."

Several native nations, each occupying a particular, though hardly precise, territory of the Northeast, joined together in the Algonquian confederation: the Micmac, Abnaki, Pequot, Narrangansett, Wampanoag, Massachusett, Pennacook, Delaware, Pocumtuck, Mahican, Wappinger, Montauk, and Delaware. Of the nations in the Algonquian confederation during the 1600s, those of southern New England had the largest populations. The most well known of these tribes was the Wampanoag, which extended its influence from the eastern shore of Narragansett Bay to the outer beaches of Cape Cod. The tribes of the Outer Cape—the Nawsets and Monomoyicks—were not directly under the leadership of the larger Wampanoag group, but historians believe that these tribes appear to have been in general accord with the leadership of the region's Algonquian tribal chief, also known as the sagamore or sachem.

By the time that European explorers reached the Americas, native peoples in New England had evolved ways of life and beliefs that rendered them healthy, strong, long-lived, skilled in basic herbal medicine, and expert at hunting, gathering, preparing, and storing food. European accounts—most notably, Champlain's logs—contain some information about the Wampanoags of Cape Cod and the nearby islands. Distinct settlements were apparent along the shores; the native people were less hunters than good fishermen and small-scale farmers. Their gardens were not limited to corn, but also contained beans, squash, Jerusalem artichokes, and tobacco.

The spirituality of the indigenous people was based on the conviction that the land was fundamental to life, and their beliefs emphasized respect for the power and majesty of nature. Their way of life was simple, primitive even, when assessed by modern standards; but it was an existence that ritualized kinship among all of nature, whether humans, plants, animals, land formations, the sea, or the weather from one time of year to another, from one place to another. These native peoples derived their identity from the land they inhabited, and they understood the intrinsic importance of place to every aspect of their existence.

Spirituality was interwoven into every facet of individual and tribal life, and cherishing nature lay at its heart. Everything in the natural world was part of the creator, who formed it and had a special, particular role. In that sense, all of life—from grass and trees to birds, fish, and humans themselves—was sacred, inviolable, and worthy of respect. Everything—hunted animals, harvested plants, trees, storms, and wind—was imbued with spirit.

By contrast, white people brought to the shores of the Americas a religious ferocity that more often than not denigrated native peoples and desecrated nature. These early settlers valued individual freedom, unfettered commerce, and ownership, none of which enhanced native life and traditions. For the indigenous tribes of the northeast, Europeans turned out to be intruders who did not share their sense of morality or a relationship to the land. While the whites may have offered new metals, fabrics, tools, and foods, they also brought grave misfortune: diseases to which the native people had no immunity.

The European presence foreshadowed a terrible fate for many thousands of Indians, first along the coast and then inland. By 1620, epidemics of what historians now believe was a flulike illness virtually wiped out populations of native people from Maine to Massachusetts.

Using Colonial sources, some historians estimate that between 1610 and 1670—when white settlers were firmly entrenched in the New World—the indigenous population had been pared by one-half. The native population in pre-Colonial times had peaked at an estimated one million across North America (750,000 in what would become the United States and 250,000 in Canada), only to be devastated as the new nation grew. By 1910, there were fewer Native Americans than ever—an estimated 250,000 across the continent. Whites—with their expansionism and sense of entitlement—had wiped out the rest and annihilated much of the native way of life.

THE EARLIEST VISITS BY WHITES to eastern North America occurred between about 1000 and 1500 A.D., depending on whose claims one accepts as valid. Records clearly indicate that Viking and continental European explorers charted the waters of the New World, "discovering" them during the late tenth and sixteenth centuries, respectively. Documents from 1523 and 1524 show the Italian adventurer Giovanni da Verrazano, under the authority of Francis I of France, voyaging along the North Atlantic shore from New York to Maine. Verrazano landed to trade with Algonquian tribes in Narragansett Bay, off Rhode Island, and passed through waters off Massachusetts—and more precisely, around Monomoy and Cape Cod—before encountering some Abenaki Indians farther north, near Casco Bay, in Maine. Verrazano's voyages gave Europeans a better notion of northern coastal America and beyond, and by 1578, more than 350 ships were said to be fishing in waters off Newfoundland.

The presence of the English off the shores of Cape Cod, and perhaps on Monomoy, was documented in 1602, when the English navigator Bartholomew Gosnold became intrigued with an area of the New World called "Northern Virginia." It was, in fact, present-day New England, and Gosnold sailed all along the territory of the northeastern coast of America, from present-day Maine to Massachusetts. One account of the voyage, kept by a crew member named John Brereton, documents the landing on the "mightie headland" many historians accept as Cape Cod. There, they made contact with native Algonquian tribes, who lived all along the Outer Cape, in Nantucket Sound, and on the Elizabeth Islands. A small amount of bartering was accomplished without incident; but because of an inadvertent skirmish with native hunters that left an Englishman wounded, none of Gosnold's crew was willing to stay behind and establish a permanent encampment in the area. They likely had met up with Wampanoags, as well.

But encounters with the native peoples were not the only memorable events Gosnold faced; the other unforgettable type of meeting was with the sea and shoals around Monomoy. In journals of the voyage, Monomoy Point is referred to as "Point Care," and those sentiments were echoed again and again by those navigating the waters off Monomoy. The French designated the shoals of Pollock Rip, off Monomoy Point, as "The Great Rip of Mallebarre." Champlain in his log described it as a "very dangerous place," which it was—and is—a shoreline vulnerable to wind and waves from the east, south, and west, as well as to tidal currents from both sea and Sound. As distinguished from "Port de Mallebarre" which is believed to have referred to what is now Nauset Harbor, not Monomoy, the Rip of Mallebarre at South Monomoy Point proved to be too great a challenge to the captain and crew. In October 1606, Champlain's ship ran aground and broke its rudder. The crew was forced to take shelter on Monomoy until the ship could be repaired and the voyage resumed.

Historical accounts from the pre-Colonial era of exploration suggest that in their canoes, the native Monomoyick Indians were expert navigators and could maneuver without trouble through every shoal, rip, and guzzle around Monomoy and the other shores of Chatham. White adventurers and ships' captains, however, were less familiar with the treacherous waters and the route of safe passage. Dutch explorers who later traversed Monomoy and the surrounding waters described them with the name "Ungeluckige Haven." And in 1620, the Pilgrims aboard the *Mayflower* were forced by the deadly waters of Mallebarre at Nauset, Pollock Rip to the south, and Monomoy Point to turn back toward Provincetown for safe landing before setting sail for Plymouth.

MORE THAN A CENTURY ELAPSED between the initial European encounters and the first recorded white settlements on the sandy peninsula of Monomoy. This is hardly surprising, given the difficulty of passage around Monomoy Point and the risks entailed in trying to land on the island's outer beach, overlooking the Atlantic. It simply made more sense to choose a harbor farther north, such as Nauset or

Provincetown. Throughout the 1600s, the entire area now known as Chatham was referred to by its Indian name, "Monomoyick," or "Monomoit." The word had so many variations, mistakenly translated in French, English, and Dutch references, that it is frequently difficult to discern which notations in early writings referred specifically to the long narrow bar of land that we designate as "Monomoy" today, and which were indicating Nauset Beach or Marsh, Pleasant Bay, or even parts of Chatham. A footnote in William C. Smith's 1947 history of Chatham signals the complicated trail the very name presents:

> The name (Monomoyick or Monomoit) is spelled in so many different ways in early records and writings that it is difficult to determine its exact form. Governor Bradford, Governor Winslow and other early authorities give the name as Manamoiak, Manamoyack, Manamoyake, Manaoycke, Monomoyick, Manamock, Manamoyick, Mannamoyk, Monomoyack, Manamoik, etc. Later, the consensus of opinion seems to have been that the final sound in the word should be nearer the sound of the letter "t" than that of the "k," and we find in the Plymouth County Records, in many early documents, and in the writings of the clergy, the following forms of the word: Mannamotte, Mannamoiett, Manamoyett, Manemoyet, Mannomoyett, Monnamoyett, Mona- moiett, Monamoit, Manamoit, Monomoit, Monnamoit, Mannomoiett, Manamoiett, Manomoytt,

> Manamoyet, Mannamoyt, etc. The forms Monumoi, Monamoy, Monnamoy, Monemoy, Manamoy, Manimoy, Monomoy, Mannamoy, Manemoy, Manamoye, Manomoy, etc., are corruptions of the Indian, which were used locally and colloquially, although sometimes found in the public records. The forms, Monnamoy and Monamoy, appear in the town records.

Whatever ambiguity exists about the exact location of Monomoy and the spelling of its name, there are things we do know about the area during the seventeenth and eighteenth centuries. Based on the recorded histories of Chatham, it is clear that from the earliest times, Monomoy—both mainland and peninsula—was regarded as land worth having, no matter how difficult it might be to reach. And for those living along its shores, there was the guarantee of plentiful food, particularly fish, shellfish, and shorebirds.

Beyond the natural and human history accounts written by early European explorers, the principal records about Monomoy tell of the long squabble over ownership of land in "Monomoit." The dispute was between the early settler William Nickerson and various governing authorities that intervened in his 1656 purchase from the Indians of substantial acreage in what is now both the town of Chatham and the Monomoy islands. Because Nickerson did not follow the regular legal course of obtaining permission from the area's then-governing body, Plymouth County Court, before he proceeded to buy land from the Indians, his ownership was not recognized—or fully realized—for another sixteen years. His agreement with the Indians was that he would deliver a small open boat, ten coats of cloth, six kettles, twelve axes, twelve hoes, twelve knives, forty shillings in wampum, a hat, and twelve shillings in money. In return, he was to gain ownership of a large tract of land in Monomoit. The contested purchase resulted in other claims being made on the land, which delayed the final settlement until 1672. Nickerson then bought up more property, until in 1682 he held clear title to four thousand acres, according to the town's history.

Most of the disputed land lay outside the current Monomoy islands' boundaries. Not much is known of human history on the then-peninsula, though there are records that the area was used in a limited way from time to time. A tavern established in 1711 at Wreck Cove, on Monomoy's west side, offered refuge to shipwrecked sailors and their passengers. In addition, the meadowlike moors extending down the center of today's islands were used in the mid-1700s for pastureland by sheep farmers from Chatham.

Permanent human settlement was uncommon—even in the limited sense of the term "permanent" as meaning "year-round." Though the constantly shifting sand mass provided access to the sea and long had been a desirable site for seemingly inexhaustible reserves of fish, seals, and whales, more navigable harbors existed to the north and the south. During the nineteenth century, however, a harbor was located at the present-day Powder Hole, near the southern end of the Nantucket Sound side of Monomoy. Known as Whitewash Village, and distinguished by its numerous whitewashed buildings nestled in the ivory-white sandscape all around, the settlement served as a harbor and fishing village for processing the mackerel, cod, and other fish taken by fishing fleets out of Chatham and shipped to New York and Boston.

Among the recorded details of village life that have survived to this day is the going wholesale price of shellfish. A fair-sized lobster brought two cents; smaller ones ran a penny. Comparable prices today are five hundred to six hundred times what they were in the late nineteenth century.

Most of the records of the village and the lives of its inhabitants were lost to two separate fires in Chatham in 1827 and 1861, but the Massachusetts Audubon Society—which has maintained a presence on the islands in various ways over the last half century—published an account of Monomoy's history that weaves together the anecdotal and existing material left about Whitewash Village. By 1839, the village included several wharves, supply stores, substantial houses, a tavern, and a school, according to Wallace and Priscilla Bailey, coauthors of *Monomoy Wilderness.* Started as a summer program, Public School No. 13 expanded to four terms and in 1860 had sixteen students.

Also located at Whitewash Village was an establishment called "Monomoy House," which was described in a February 1865 article in *Harpers New Monthly Magazine* as "a barracky, amphibious structure, fishermen and coasters' fitting-store on the first floor, lodging-house and excursionists' inn on the second." It also housed scattered storage and packing sheds. The inn was characterized as nearly "amphibious," because occasionally in winter tides were so high they would lap up the front stairs. During such swells, boys and girls would use a gondolalike boat to go to and from school. And on occasions when the boat was not available, the boys would carry the girls, wading their way to class.

But even with the village's pluck and tenacity, the sea, storms, and sand as ever held the community's fate in the balance. Their action was predictable, and for the harbor, irreversible. Shortly after 1860, brutal storms already had reconfigured the scheme of things in Whitewash Village and in the harbor itself. Sand, driven by surf around the hook of Monomoy Point, drifted over the opening to the harbor, filling it and effectively closing it to larger vessels. The village's heyday, from 1830 to 1860, came to an end, but summer residents continued to inhabit the area well into the 1930s. The last cottage on Monomoy rested at the south side of the present-day Powder Hole until the close of the century.

Other parts of the island accommodated human presence in various ways. For seventy years, from 1862 to 1932, the place bore the imprint of sportsmen who swarmed to Shooters Island and hunted from the shanties of the Monomoy Branting Club (named for the brant, a small black-necked goose about the size of a mallard). The organization, which allowed members to use the hunting camps for a week at a time in fall and winter, maintained "sink boxes" placed in holes on the outer edges of marsh along what is now Hospital Pond on the northern edge of South Monomoy. The boxes were large enough to provide cover for two people—but only while the tide was out. When the water level reached high tide, the boxes literally would sink beneath the surface, and the hunters would head back to the club for shelter and dining.

During a time when hunting was a popular accepted pastime, the island epitomized a gunner's and hunter's dream. But by the 1930s, bird populations moving on and over the Outer Cape's barrier islands were decimated by the sport. Massachusetts Audubon estimated that in the late 1800s, as many as 8,000 golden plover and Eskimo curlew were shot in a single day on Monomoy. Even today, several of the names of ponds and other land formations bear the mark of the hunting era. Hospital Pond was so named because wounded ducks would retreat there for refuge or, more often, to die.

But the graveyard most often associated with Monomoy was the sea. The surrounding shoals and rips were so great a peril that an estimated three thousand ships and boats were wrecked along its shores in the past three hundred years. And for mariners unfamiliar with the hazards, or inattentive to how fast the offshore picture could change, danger could be deadly. There was no avoiding the possibility of trouble; the only hope lay in getting expert guidance, ample warning, and ready help. That came only with the Monomoy Light.

Seasons of Light

You have to go out, but you don't have to come back.

—Motto of the surfmen of the
U.S. Life-saving Service

NOTHING IS DARKER THAN THE SEA BY NIGHT and nothing more dangerous than its hidden shoals.

Even in broad daylight, the most experienced mariner can be overwhelmed by trouble along Monomoy's shores. The engine's blades churn in sand, the bow is buffeted against the bars by the insistent rips, and the surf pounds toward shore. Faced with these ordinary conditions—let alone a gale—a captain's only sensible plan is to try to keep from running aground. Or worse, breaking apart.

It's always been this way off Monomoy, considered by many to be the most deadly waters along the New England coast. From the earliest landings until today, ships' captains have required assistance to navigate safely around the islands. What they needed was a marker warning of dangerously shallow waters, some way to fathom their position, an illumination in the dark.

For nearly two hundred years, the guidance was anecdotal at best. There were rudimentary maps and reports from friendly Indians whose directions, in an unfamiliar language, might or might not be understood. Often the best available guide was a mariner's own instincts, common sense, and experience. All that was worth something, to be sure, but to negotiate the seas off Monomoy, it frequently was not enough. A sailing vessel of the early seventeenth century could not execute sudden changes of direction or speed to evade hazardous low water and sandbars on a moment's notice. The captain and crew of the *Sparrowhawk* learned that too late, when, one fall night in 1626, as they were embarking on the last leg of their voyage to Virginia, they ran aground south of old Monomoyick Bay (now Pleasant Bay). The following morning, as the tide ebbed, the crew and passengers were able to wade to shore, but, according to accounts of the incident, they had no idea where they were or what they should do next to save the ship.

Wampanoags paddling through the bay stopped to help and offered to take two of the crew to Plymouth, where the men enlisted the help of William Bradford and his crew. The party returned to the site and repaired the ship, only to learn days later that the *Sparrowhawk* had barely set off before running aground again, as a gale hit. After the storm, the ship was no longer seaworthy, so it was abandoned, and the passengers were taken to Plymouth for the winter. The ship, completely at the mercy of the elements, was soon buried by sand. Over the next 250 years, it would be repeatedly disgorged and reburied by the sea, until its remains could be salvaged. A hurricane uncovered the *Sparrowhawk* briefly in 1782, but it was quickly covered by sand and remained hidden until, in 1863, a gale exposed the ship again. The hulk, by then an artifact, was taken to the mainland.

The small English ship was the first in a long line of vessels claimed by the sea. Since 1600, an estimated 3,000 ships have been wrecked off Cape Cod, many along the ocean shoreline. The calamities prompted people to help in whatever way they could, and lifesaving efforts evolved along with tragedy along the New England coast. Once Europeans began settling on Cape Cod, it was only a matter of time and human ingenuity before lifesaving would become an organized enterprise along the outer beaches and around the offshore islands. And on Monomoy, striving to prevent disasters and offering succor to survivors have been basic facts of life for nearly three hundred years; first, through the lifesaving stations, which responded to foundering or wrecked vessels; and second, through lighthouses and lightships, designed to guide ships through dangerous waters.

Bradford's aid to the *Sparrowhawk* foreshadowed the work of the U.S. Life-Saving Service, an early branch of the Coast Guard; but the real origins of publicly organized lifesaving crews can be traced back to China. In 1708, the Chinese developed a network of manned stations stocked with lifesaving boats outfitted for rescues along the Yangtze River. The English and Dutch began organized lifesaving in the second half of the eighteenth century, and naval historians believe that trade with Asia by both Britain and Holland led to similarities among the formal rescue efforts of all these nations.

In New England, especially in Massachusetts, lifesaving measures developed almost simultaneously with the building of lighthouses along the Atlantic. The first lighthouse was erected at the entrance to Boston Harbor; its first "fog horn" was a gun blast. The light's completion was welcome news, and an announcement of its operation appeared in the *Boston News Letter* on September 17, 1716: "Boston. By virtue of an

Act of Assembly made in the First Year of His Majesty's Reign, For Building and Maintaining a Light House upon the Great Brewster (called Beacon-Island) at the Entrance of the Harbour of Boston, in order to prevent the loss of the Lives and Estates of His Majesty's Subjects; The said light House has been built; and on Friday last the 14th Currant the Light was kindled, which will be very useful for all Vessels going out and coming in to the Harbour of Boston, or any other Harbours in the Massachusetts Bay, for which all Masters shall pay to the Receiver of Impost, one penny per Ton Inwards and another Penny Outwards, except Coasters, who are to pay Two Shillings each, at their clearance Out, And all Fishing Vessels, Wood Sloops, etc. Five Shillings each by the year."

The Boston light soon was joined by sixteen other Colonial lights scattered along the Atlantic, from Maine to Georgia. Most were still operating in the twentieth century.

By the 1780s, only shortly after the Revolutionary War, rescue efforts in Massachusetts were being organized by the Humane Society, which built small huts to house shipwrecked sailors and passengers. Survivors who made it to shore could find necessary provisions—food, candles, kindling, and fuel—to hold them until other help came. But the huts did not have boats to assist in the rescue of those still stranded aboard a sinking vessel; and vandalism and theft of supplies in the shelters were constant problems.

Lifesaving boats were not used until the nineteenth century, when a Nantucket boatbuilder developed a specialized craft that was essentially a whaleboat that relied for power on ten rowers and steering oars. By the late 1800s, the Humane Society had added seventy-six lifeboat stations to eight existing huts on the Massachusetts coast. By then, the stations were generally stocked with boats, rafts, and line-throwing devices and were manned by volunteers. But they covered Massachusetts only, and all the maritime states needed rescue networks. It was not until the 1870s that the federal government was prompted into action by the catastrophes of several shipwrecks. By 1879, the responsibility of caretaking the stations was assigned to the new U.S. Life-Saving Service, and within a matter of years, stations were manned by full-time keepers and crews of surfmen, who were paid—however poorly. By the turn of the century, keepers were earning $900 a year (up from the original $200 annual salary allocated to them). Surfmen were paid $65 a month.

Lightships were introduced in 1819 off the coast of Virginia, and Congress funded New York Harbor's first lightship in 1822. By 1909, there were fifty-six lightship stations along the U.S. coastline. The vessels were essentially floating lighthouses, often manned by crews but occasionally left unattended and merely anchored offshore. For decades the ships were so neglected that an 1852 federal report found that "nothing could be much worse than the floating lights of the United States." Because the early lightships often were held in place with only one anchor, the vessels frequently turned into hazards themselves when torn from their proper position and left to drift.

The nineteenth and early twentieth centuries brought spectacular changes in shipping and rescuing. Management of, and funding for, the Life-Saving Service had been bounced from one federal agency to another and back again. But by the start of World War I, the service had thirteen formal districts on the nation's coasts, lakes, and rivers and 271 stations, plus eight houses of refuge between South Carolina and eastern Florida. The Lighthouse Board, which assumed control of the nation's expanding and still-derelict system of lighthouses and lightships, was established in 1852, and in 1910 was reorganized within the Department of Commerce as the Lighthouse Service under the Bureau of Lighthouses. In 1939, the U.S. Coast Guard took over the service, turning control of the nation's maritime lights back to the Department of the Treasury. The service oversaw nearly 1,400 lights, 54 lightships, and 225 light buoys.

All the Cape's lighthouses were built during a period of about seventy-five years, beginning in 1796, when Highland Light was erected on the cliffs of North Truro. Still others were needed: the Twin Lights of Chatham in 1808, Race Point in 1816, Monomoy in 1823, Long Point in 1826, the Three Sisters of Nauset in 1838, and Wood End in 1872. By then, nine lifesaving stations also had been built along Cape Cod, including one on Monomoy. By 1900, a second was added at Monomoy Point, and by the 1930s, Cape Cod was safeguarded by twelve lightships and nineteen lifesaving stations, counting six on the offshore islands.

As many as four lifesaving stations at a time watched the seas between Chatham and Monomoy Point. Additionally, eight lightship positions were maintained around Monomoy Point throughout the latter half of the nineteenth century—Pollock Rip Shoals/Slue, from 1902 to 1923; Pollock Rip Lightship, at three distinct locations from 1949 to 1969; Stonehorse, at two different locations near the point, from 1916 to 1963; Shovelful Shoal from 1852 to 1916; and Handkerchief, south of the point, from 1858 to 1951. Lightships often had to change positions as the bars and flats off Monomoy shifted, and often they were pulled off their mark by storms and high seas. Pollock Rip Lightship was tossed around by the elements so frequently and capriciously that it became known by seamen around the Cape as "The Happy Wanderer."

Monomoy Light was commissioned by the federal government less than fifty years after the birth of the nation and was built on four acres of land on the island's southern beach. The forty-foot cast-iron tower lined in brick was lighted five years later, in 1828, fired by oil. It was manned for nearly a hundred years, but after 1923, the more powerful lights at Chatham and on Nantucket made its operation redundant. Upkeep of the extinguished light was ignored for many years, but the structure was restored in 1988 and still stands today.

After its decommissioning, the light and the two-story, wood-frame keeper's house passed through a number of private owners, including the Massachusetts Audubon Society, which bought the lighthouse complex and two surrounding acres in 1964 and held them until the mid-1970s, when the buildings were taken over by the federal government. Today the light, the keeper's house, and a fuel storage shed serve as a center for ornithological research and natural- and cultural-history tours run by the Cape Cod Museum of Natural History, in Brewster. Audubon's Wellfleet Bay sanctuary, too, continues to offer birding tours of the island when weather and sea conditions permit.

Lighthouses and lightships rendered the New England shoreline safer, but not free from risk. There was no changing the sea's erratic moods or protecting unskilled captains and crews from their own folly. Before the advent of radar, sonar, and other technological maritime advances, it was almost impossible to travel along the seacoast without encountering one peril after the other, and the waters around Monomoy were notorious for irregular depths and crosscurrents. Even contemporary offshore-depth maps illustrate the chaos of depths and shoals around the island and at the confluence of Nantucket Sound and the Atlantic. The general outline of the maps shows a fairly regular pattern, like fine lace, spreading out from the Cape Cod coastline; but at Monomoy's Pollock Rip, it is as if a great snag were tangling the design. That snarl was exactly what mariners had to worry about and some settled for a safer, more time-consuming, wide sweep into the Atlantic to avoid the choppy, low waters off the point.

Most of the serious shipwrecks off Monomoy occurred during heavy nor'easters or under hurricane-force winds as vessels tried to clear the confusion of shoals. Sitting safely on land, one cannot imagine the hellish tumult of the storms or the risks that lifesavers faced during rescues, but the catalog of doomed ships off Monomoy gives some indication. A twentieth-century map of marine disasters off Cape Cod shows the shoreline dotted with hundreds of wrecked vessels and lists more than 125 ships lost off Monomoy between 1800 and 1950. Wreck reports of Monomoy Life-Saving Station from late September 1883 to mid-November of 1887 chronicle rescue attempts for nearly seventy ships.

The surfmen of the U.S. Life-Saving Service risked their own lives every time they put out in their boats for a rescue. Their vessels, measuring from twenty-four to thirty-six feet in length, commonly were buffeted by gale winds and waves as high as a three- or four-story building. Service regulations compelled station keepers and crews to attempt rescues, no matter how overwhelming the odds of success or even of safe return. So routine and accepted was the peril that the surfmen's stoical motto became, "You have to go out, but you don't have to come back."

And many did not return.

Of the scores of heroic rescues that took place off Monomoy, two—separated by fifty years—stand out: the 1902 wreck of the coal barge *Wadena* and the 1952 disaster involving the tanker *Pendleton*. The tragedy of the *Wadena* epitomized why Monomoy for two centuries had been called "the graveyard of the Atlantic." In one storm, on one day, seven of nine surfmen from the Monomoy station and five men aboard the stranded barge died trying to get the vessel to shore and themselves to safety.

The *Wadena* was one of two coal barges being pulled by tug from Norfolk to Boston when a vicious northeast gale struck the evening of March 10, 1902. Just after midnight, the storm had become so ferocious that all three boats were in trouble; about one o'clock in the morning, the barge *Fitzpatrick* ran aground off Monomoy Point, and the *Wadena* hit Shovelful Shoal just east of the point. Neither seemed in immediate danger because each carried such a heavy load that it settled on the bars. Lifesavers from the Monomoy Station rowed out and took ashore all but a few of the men aboard the two boats. The tug steamed on to Hyannis to report the damage to the owner of the two barges, William H. Mack, of Cleveland, Ohio, who by chance happened to be in town. Mack organized a crew of "wreckers," or salvagers, to shovel the coal off the barges to free them up, and he hired two more tugs from Boston to tow the vessels to port.

The tugs made it to Monomoy Point on March 16, but before they could retrieve the barges, a second storm blew up from the southeast and delayed the salvage operation. Most of the wreckers and crew returned to the mainland aboard one of the tugs, but five men—Mack, the *Wadena*'s captain, steward, and two others—stayed aboard to ride out the storm. Just before midnight, the gale winds intensified and storm swells poured across the barge, washing over the deck time and time again. The barge lifted and crashed repeatedly on the hard bottom of the shoals, and by 3 A.M., it began splitting under the punishment of the storm and seas. Mack, attempting to signal for help, raised an American flag, upside-down, on the rigging. But the storm had fogged in the island, and nearly six hours passed before the keeper of the Monomoy Station, Marshall Eldredge, who was walking the shoreline to watch for the *Wadena,* saw the banner. He rushed to the telephone shanty and called seven surfmen to help. While the crew put their lifesaving boat into the rough seas, Eldredge walked and waded as far as he could so that he could join them.

The surfmen struggled out to the *Wadena,* where they were able to get all but the steward into their boat. At first too terrified even to move, he finally jumped from the deck, but did not clear the lifeboat. He crashed aboard, smashing through the wooden bench-seat in the stern. The boat, now damaged, held thirteen men, and was immediately engulfed. The men from the *Wadena* became so frightened that they clung to the surfmen; and as the assault of waves continued, the boat capsized, throwing all the men into the churning water. Those who did not succumb immediately held onto the overturned lifeboat, but all except one were drowned. The sole survivor was the surfman Seth Ellis of Harwich Port, who watched his station captain and crew slip off, one by one, into the stormy seas. He reported the dismal, last words of his comrade Arthur Rogers, a surfman from the Monomoy Station: "Rogers had lost his strength," he recounted. "And failing to get a more secure place on the bottom of the boat, feebly moaning 'I have got to go,' he fell off the boat and sank beneath the waters." Ellis managed to hang on to the overturned lifeboat until he was saved in a daring rescue by Elmer Mayo, a local carpenter and wrecker who was aboard the *Fitzpatrick.*

Through the bitter, late-winter storm and heavy fog, Mayo had spotted the disabled boat and Ellis holding on. Though he himself could not swim, he decided to attempt a rescue. Over the protests of the *Fitzpatrick*'s captain, who did not think it possible to reach the lifeboat, Mayo resolved to launch a makeshift lifeboat—a dory and two oars, one long and one short—for which he fashioned thole pins to attach the oars to the boat so that he could row the small craft. Despite the mid-March cold, he stripped to his underwear and set out for the overturned boat. He was almost overcome with exhaustion before he made it to Ellis's side, accomplished the rescue, and got the dory safely to shore. Even so, twelve men—including station-keeper Eldredge and six surfmen—had lost their lives.

In the wake of the disaster, questions were raised about whether life belts had been stowed aboard the lifesaving boat and whether they would have made a difference in the outcome of the calamity. Ellis, who was named keeper of the Monomoy Station after the tragedy, made no mention of the belts in his account and blamed the disaster on the terrified men the lifesavers had attempted to rescue from the *Wadena:*

"As I have often said, if the persons we took off the barge had kept quiet as we told them to, all hands would have been landed in safety."

For his part, Mayo was hailed as a local hero and was awarded the Congressional Medal of Honor and the Massachusetts Humane Society's Gold Medal for bravery.

Fifty years later, two more tankers, the *Fort Mercer* and the *Pendleton,* broke in pieces east of Cape Cod during a midwinter gale that lasted two days. Aboard the two vessels, which were only forty miles apart when they sustained their final damage, were eighty-four men. By the storm's end, seventy had been saved; fourteen were lost.

The first distress call came on the evening of February 18, 1952, from the *Fort Mercer,* which had been torn in two south of Chatham by sixty-foot seas and seventy-knot winds. Coast Guard cutters responded from several stations along the northeast coast. Rescuers had difficulty even locating the sinking hulks in the tumultuous storm, but aided by a plane sent from the Salem station, they managed over the next day to save twenty-seven men. Thirteen died.

As the drama was unfolding that evening aboard the dashed stern and bow sections of the *Fort Mercer,* the Chatham Coast Guard Station commander learned of another impending tragedy. The *Pendleton,* en route to Boston that morning, had broken up off Provincetown, disabling the radio, so that a distress call was impossible to make. It was not until wreckage from the tanker washed up on the shores of Truro that the Chatham station was alerted to the crew's plight. For twelve hours, the remains of the boat had been buffeted about erratically, drifting south until they were off Monomoy. With so many rescuers focused on the *Fort Mercer,* there was no alternative but to send men out from the

Chatham station at the height of the storm, across the deadly channel bar.

A twenty-three-year-old coxswain, Bernard Webber, was placed in charge of the rescue attempt, supported by three other guardsmen. Together they braved the gale and seas that Webber later described as a force that "no human being could come through . . . on his own doing." They set out in a thirty-six-foot motorized lifeboat, but before they reached the *Pendleton,* the surf had broken their windshield, and the compass and some lifesaving equipment was lost overboard. The engine failed temporarily; Webber got it to start again by smacking the carburetor with a hammer.

They plowed on, at last discovering the stern section of the tanker and all thirty-three surviving members of the crew clinging to it. Miraculously, the guardsmen were able to navigate past the sinking stern over and over again, rolling in sixty-foot waves, without colliding with the hulk. Each time they passed, a few more men would jump to safety. Webber realized midway through the rescue that the lifeboat, which could safely hold a maximum of twenty-three passengers, probably would not be able to carry all of the *Pendleton* crew. But in that moment, he was determined not to leave anyone behind. In the furious seas, only one man was lost, when he became trapped between the tanker's rudder blades and the lifesaving boat and was drowned.

Even with the survivors aboard at last, the riskiest part of the rescue still lay ahead—getting back to safety by crash-landing on Monomoy. Because their navigational equipment had been swept overboard, the men had to guess where they were and where they ought to be headed. A Coast Guard cutter some distance offshore radioed to Webber that the lifeboat should head away from Monomoy and out to sea so that the men might be gathered up by the cutter. After listening to a debate among several Coast Guard officials about the wisdom of such a move, particularly with an overcrowded boat, Webber simply shut the radio off and steered for Monomoy's shore.

The boat had been out in the storm for five hours, and the men were not at all sure where land was. When a landmark finally appeared, it turned out to be the flashing red light of the buoy at the entrance to Chatham Harbor—miles from the wrecked tanker stern and from the southern edge of Monomoy Island. The boat and everyone aboard made it safely into the harbor.

Webber and his crew were awarded the Coast Guard's highest honor for bravery, the Gold Lifesaving Medal; but the coxswain never considered his actions or decisiveness to be heroic. "Even if it meant death," Webber said, "you did what you had signed up to do."

The hulk of the *Pendleton* remained visible offshore until the blizzard of 1978, when high winds and seas again set the remains of the vessel adrift. Without a stable location, the ghost tanker stern became a serious hazard, and it was dynamited and sunk the following summer.

NO SIMPLE RECITATION of the historical facts of the lights—and lifesaving—in and around Monomoy can do justice to the way of life surrounding these early sentinels of the sea. For the lifesaving surfmen and the lighthouse keepers, as well as their families, daily routine could turn to chaos in a matter of hours, or even moments. Good weather might bring a blissful monotony on quiet days; but storms could just as easily foment the seas to fury.

Of course, public and private accounts of Monomoy's history are full of idyllic scenes and nostalgia, particularly from records kept in summer at the light or at the Powder Hole harbor. There were no hard-and-fast rules for the lighthouse keepers' families, but some wives and children shared in the keeper's solitary duty by living for several months of the year at the light. One keeper, Asa L. Jones, resided at the brick tower and cottage from 1875 to 1886—peak years of the light's service. His wife, Clara Freeman Paine; son, Maro Beath; and their dog, Spot, lived there during summer and returned home to Harwich for the winter, leaving Jones to endure alone the bitter cold and loneliness of the winter light.

In 1884, Maro, then nine years old, was encouraged by his father to keep a journal of his life at the Monomoy lighthouse. His diary entries were not the only impressions of island life that he recorded. Long after he and his family returned to the mainland, the island light guided him through the tides that would shape his future. He grew up and became a language professor, his interest piqued by reading the instructions to equipment in the lifesaving station, which were written in French, German, and Spanish. Maro went on to write a short story entitled "Monomoy Light Adventure," which, though unpublished, remains a vivid rendering of a keeper's and lifesaving surfman's existence. In it, he sketched characters whose portraits are drawn from the people he had met and known during his summers on Monomoy, and the landscape of the island is as animate as its residents.

> This was the hour of sunset, the hour when the lamps of all the lighthouses and all the lightships . . . are lighted for the service of night. When days are not clear and the sun cannot announce its own departure from the horizon, the keeper has to rely on his almanac to know the moment when he must strike his match and kindle the little wick that takes the place of sun and guides the ships by night from the perilous reef and shoal. At this twilight hour vision can play weird tricks, and when the mist and drizzle are closing in, water and sky meet, or rather, they become as one, and under such conditions it is extremely difficult to distinguish and locate objects at sea.

Jones's story, along with excerpts from his journal, captures a way of life that could seem equal parts natural playground and prison. Even in summer, when weather was generally more serene, the isolation of island life could grow tedious. The boy's journal suggests that the family returned to Monomoy in late March in 1884 and lingered until November. They returned the following summer, apparently Jones's last before his service as keeper was completed. The diary is full of the details of a simple and now-vanished era, in which the concerns of the day were weather, chores, and occasional visitors, as well as hunting and fishing.

"Rainy," starts an entry for April 5, 1884, followed by "Good weather" on the 6th. As summer approaches, the boy's succinct evaluations run from "good day" to "very good day" to "very pleasant day" by May 16, when the most notable event was that "Papa painted his lantern." By June 17, the weather had turned "awful hot," he notes, and even in late August, the weather was still "very warm." The last entry of the season reads: "November 28. We came home and left Papa all alone."

The Monomoy Light was decommissioned by the federal government in 1923, a century after it had been built. Years after he had recorded his experiences on Monomoy in his journals, Jones reread his entries and penned commentary into the margin of his notes of September 16, 1885. He was saying good-bye to the light and his family's time there. "Pitiful," he wrote. "On July 30, 1925, the lighthouse was sold."

SOMETIMES IN WINTER I still long to go and stand in the shell.

I find myself, in the cold, taking warmth from the recollections of the barrier beach. While I hibernate on the mainland, the squalls carry on, the wind whines without ceasing, and the surf recasts the wilderness of Monomoy, brought forth by little more, it seems, than sand and stubborn intention.

I imagine all the places I might be going, all the sites I might be exploring and marking, if it were a different season—by the calendar and the drift of my life. Of course, I know that spring will come around again, after the owls court and I awaken again from the cold sleep locking the land in frost and stupor. But for now, I find myself restless in odd moments, wondering about the island, fretting about the light.

The lighthouse and keeper's cottage on South Monomoy are, like shore debris on the roughly five-mile-long stretch of island, shells—remains of places other people once called home. Persistent mice and spiders oversee the rooms as empty and damp as caves. Paint peels like scrolls from the ceilings, and garter snakes burrow under the back steps. Everything, and everyone, else seems to have left, long, long ago.

One day, after the year had made a full round since my last winter walk on Monomoy during a predictable January thaw, it seemed as though Hillary and I might be able to get out to the islands for a while. An unexpected, balmy few days late in the month seemed promising for a voyage.

The night before, I could hardly sleep. I always feel the anticipatory fussing of a child when I know there's a chance to get away to the light, and I woke before five, fully an hour and a half before I was to get a wake-up call or to set off for the pier. Outside it was twenty degrees. I knew from that piece of news alone that the trip wouldn't come off, but I got up anyway, made coffee and a fire, and padded around the house, anxious for the call. At 6:30 the phone jangled in the front room, startling the dog and halting my circle through the house. It was the skiff's captain, reporting simply: "It doesn't look good." We agreed to try for another time in a month or so and rang off.

I hovered in the front room to tend the fire and keep warm, and the dog nosed her way over to my side, angling for a walk. But I wasn't ready to go. She flopped down with a resigned sigh and rolled onto her side, her back to me. Settling down some seemed a good idea, so I joined her, using her resentful back as a pillow; and I half-dozed, half-meditated, conjuring Monomoy even though I couldn't make the shore that day. Instead, I imagined the trip, drawing on all the detail stored by the senses from a dozen voyages made before—the rocking skiff, the rolling seas and spray, the matted wrack, the white sands and grasses gone from green to silver in the sun. I walked the old path, the long way, around the tidal pond, past the Poirier camp, the egrets studying the shallows and keeping an eye on me. I moved along the scant path, along the spine of the island to the light, to the boarded-up cottage.

In my mind I always have a way inside. Even now I can tumble all the combination locks in the proper sequences to guarantee a way in, to stand inside the husk of the house where keepers of the light took shelter with their families, while the great beam guided ships around the point, and on farther, till the light of Nantucket took over. Now the place—still stocked as it was a century ago, with coffee, cans of milk, soup and spaghetti sauce from a summer of birders winging through, the rusty pump needing a prime—looks for all the world like an oversized lifesaving station, equipped with all the essentials of living until human help arrives. But, of course, the lifesaving stations along Monomoy are long gone, too; and anyone deserting the wreck of a boat and scaling the dunes to get to the light likely would find no aid there. The only long-term occupants now are mice, bats, wilderness, and memories.

When spring comes early to the island, as the birds return, I will go back to the edge of the Powder Hole, and I will take to the trail, watching for the ghosts of Whitewash Village and the signs of the great cycles of life—the birds, the plants, the washed-out whelk and razor clams decorating the beach. In the end, I will go and stand on the deck of the light, just outside the shell, and listen to the rush of the sea and the soft wind. But I will not open this private space that belongs to more than me. I will leave it to those who keep it now—the elements, the island, and the creatures there. I will not enter, though I know the way.

Roots and Remnants

Just to be is a blessing.
Just to live is holy.

—Abraham Heschel

THEY SAY THE SACRED HEART OF JESUS kept the last cottage on Monomoy standing.

But it was the hand of man, trying to re-create a wilderness, that would take it down.

If you could see the cottage—built on sand of tarpaper and wood with an old barn door for a roof; with a tattered print of the Sacred Heart, wrapped in clear plastic and hung over the shack door frame—you'd be tempted to believe divine intervention must account for its longevity. It's hard to imagine that for more than sixty-five years, this structure could withstand the buffeting of winds and flooding from spring tides filling the Powder Hole on South Monomoy—but it has endured that, and more. The last cottage on Monomoy was the one island building to survive even the vandals.

The exact number of camps and makeshift cottages that existed over the years on Monomoy is not known. However, in 1944, when the federal government took control of the island for use as a national wildlife refuge, twenty-two camps remained, including several at the

Powder Hole. The assault of weather and wind, coupled with government regulation, took down nine more by 1970. But through it all, the cottage of Diana Poirier, of New Bedford, nestled in front of the dunes at the water's edge, held its place. It survived the blizzard of 1978 that split Monomoy in two. It weathered hurricanes; stood against devastating northeasterly storms and the inconstant terrain a barrier beach builds on; survived the risk of fire, particularly in summer and long stretches of the year when the camp was uninhabited; and remained unharmed by troublemakers who routinely damaged the lighthouse and keeper's cottage a half mile off. Every single hazard that spelled the end for other private habitations on Monomoy left the Poirier place alone.

Even officials of the Fish and Wildlife Service, who had to contend with maintenance and repairs wrought by intruders at the light, found the tenacity of the makeshift shack remarkable. Former refuge manager Ed Moses of the Fish and Wildlife Service considered it "a very interesting study in human psychology," because vandals who regularly damaged the government's locked and boarded buildings seemed to leave the rugged camp undisturbed. And the Poirier cottage was more accessible; in nearly seven decades it was never locked.

"No, no sense in doing that," says Robert Poirier, sixty-six, the son of the late owner. Instead, at the last shack, security was by the honor system: Visitors could come and go and were free to use the cottage. On the door, the simple, handwritten request, scratched in pencil into the wood, for years read: "Please Leave This Old Shack The Way You Found It. It May Not Be Much, But It Brings Much Pleasure To Those

Of Us That Have Stayed Here In Its Humble Walls. Thank You. Winter Solitude Seeker."

But by midsummer of 1999, when it became clear that the cottage was to be dismantled within a year by federal officials, a new announcement was scribbled: "PLEASE!! Leave it the way you found it, or better. It's all we have before Big Brother takes it." It was signed: "PHD, Jr. I'll be back."

Anger and bravado aside, the planned demolition of the cottage signaled a break in a cherished tradition not only for a family but for a whole community of human drifters who had sought shelter at the Poirier camp. Monomoy, so long a haven for people escaping the pace of mainland life or the unanticipated peril of stormed-churned seas, was being returned to the creatures: the tide of unfettered time in a wilderness place. But it took the authority of the federal government to ensure it.

"The Special Use Permit (#64571), dated March 17, 1999, issued to your mother Diana Poirier for personal use of Building #17 on South Monomoy Island, Monomoy National Wildlife Refuge is terminated as a result of her death, which occurred December 6, 1998," read part of the June 29, 1999, letter Poirier received from the Fish and Wildlife Service. "The permit originally issued solely to Mrs. Poirier was for her personal use. The permit is not transferable. You are advised that, with the termination of the permit, you are no longer allowed to use the property."

"It's the end of an era; it's the end," says Joan Poirier Krouzek, Diana's granddaughter and the daughter Poirier always called "the

Monomoy girl" because she was so attached to the island. "To not be able to be there, it's like losing the family homestead. It's like losing the family farm." Added to the grief over her grandmother's death, the loss of the Monomoy camp was almost more than she could bear.

"She was a woman who had very little material wealth, but this was her legacy and my grandfather's legacy to their children and their children's children," says Krouzek, who has taken her two daughters and a son to Monomoy whenever they could get away during the camp's last decade.

The gift the Monomoy legacy brought to Poirier's extended family was the grace and dignity of simple living. "The cottage is so incredibly humble," Krouzek recalls. "It teaches you that you don't need a lot to be happy. I never kept the biggest fish I caught on Monomoy. I kept the one that met the need of the day and threw the rest back. You didn't take more than you needed, and you learned 'The Way,' that connection with earth, and ocean, and nature."

It came as no surprise that the government would take control of the Poirier camp; it had owned the cottage since the midforties. Diana Poirier, like other property owners at the time, was granted a life-term use lease, renewed annually for $2.50. Nobody really thought the Poirier place would be spared the fate of all the other once-private cabins. But still, when the time came, the family was not ready.

"She chose life, always, and happiness," says Kruzek. "She lived a hard life, and she never let anything defeat her, ever."

The Poirier family acquired the camp in the early 1930s, when the late Philibert Poirier, a Canadian-born lineman, purchased the site for thirty dollars for a hunting camp—"which at that time was two weeks' pay," says Robert Poirier, retired manager of the savings department of the New Bedford Institution for Savings in New Bedford. "It would probably be a thousand now."

A few years after Philibert bought the property, he was diagnosed with Hodgkin's disease. He died in 1936, leaving his wife, at age thirty-two, with five young children. But every summer, Mrs. Poirier and the children made the voyage to the island camp—no small venture for a young widow with a large family. But she did not complain, her son says. "She swore the cabin was being held up by that Sacred Heart, and I can't deny it."

The family made the trip, frequently with the assistance of Ralph Stello, then of Franklin, who often transported the group by car from New Bedford to Chatham and then on to the camp. In those days people could still drive at low tide from the mainland to the island, and other times could navigate boats right into the Powder Hole. Travel then as now was determined by the whim of weather and the condition of the island.

Mrs. Poirier, who was interviewed a few years before her death at her home, recalled the days when motor vehicles still were allowed on the island. Days for the children were filled with sun and swimming, playing in the dunes, and monitoring the schedules of the Coast Guard lifesaving crews stationed on Monomoy. On rainy days, Mrs. Poirier remembered, the young ones amused themselves putting together puzzles and, once, making French fries—a project that took all day and attracted stragglers from other camps, too.

Above all, though, she recounted how she would watch for visitors late into the night in summer.

"I'd be sitting on my porch, and all of a sudden I'd see a light . . . see a light . . . see a light," she said, making a waving motion with her hand to simulate the movement of car lights bobbing up and down in the dunes. "And I'd know someone was coming."

The tale took on the status of myth to the Poiriers, as oft-repeated stories do in families, says Kruzek. "In my mind I can see her, my grandmother, on the porch in her rocking chair, the way she would lift her hand. I can freeze time and hear her telling that story, and I think she told it every year."

Break 1998

Whoever looks round
Sees Eternity there.

—John Clare

THE END ROSE AND FELL like a tide. Like day into night. Like the descent of rain or snow, sun or surf.

Another score of years had slipped from the human calendar, another life had come full circle, the end and the beginning balanced—light into darkness, yielding once more to light. On December 6, 1998, Diana Poirier died, without struggle, at last, at age ninety-four, migrating to unfamiliar shores. And with that finality—the flight of breath itself—the private human holdings on Monomoy came to an end. Now, though the lighthouse and keeper's cottage would remain, like memories or conquerable history, the island refuge would return to its earliest inhabitants, creatures that comprehended the sand, sea, and air in ways we still were learning.

It would be more than a year before the traces of her family's habitation, a simple camp on the shallow, empty inlet at the Powder Hole, would be erased. Meanwhile, the great black-backed gulls still held court there, pacing on the tarpaper roof, patrolling their high place of dominion over the little space humans had held for so brief a time.

And in the tidal harbor, egrets and oystercatchers, willets and sandpipers came and went, no less for not being seen. The horned larks elaborated the island moors, the harrier and great horned owl haunted the dunes and marshes. And overhead, the glint of sun on the wings of terns, the flare of stars would be the new lights to lift and dip—like so many headlamps on an old truck taking the dunes in turn, arriving home.

Still, she departed, a long-distance survivor leaving more than a perishable shack, more than a well-worn shell, in the wake of the years. Forgiveness, acceptance, simplicity, and love that knew only welcome—these illuminations her offspring carried of her, as they carried on.

And on Monomoy, the land returned to itself, the creatures, stirring, to their place.

A Prompting of Wings

How shall I name you, immortal, mild, proud shadows?
I only know that all we know comes from you,
And that you come from Eden on flying feet.

—William Butler Yeats, "The Shadowy Waters"

In autumn, the skies over Monomoy are seemingly never emptied, the winds are never stilled. August drifts into September, and the season turns, like thought into dream, summer's stir slowing into fall. But there is a restive air over the islands. Like tides, the birds move in waves on and over the sand.

By the human calendar, it is barely September, and autumn equinox floats some three weeks off. But the natural evidence of fall has been showing for weeks on South Monomoy. At dawn on one clear day, in the pond and sedge of Lighthouse Marsh, a pair of snowy egrets, two northern harriers, and four Canada geese share water, protective cover, and the search for food—at least for a time. In the poison ivy and bayberry thickets nearby, dozens of warblers and sparrows flit and peek. Red-breasted nuthatches peck for seeds in the cones on a few pines around the lighthouse, dangling like ornaments while they work their feeding-ground-in-midair.

At dusk, a mile from the tip of South Monomoy, the gulls gather.

They call insistently, scores shrieking till hundreds have arrived, their white and sooty forms rising and drifting like ash in the wind. The snowy egrets, gone at midday, make their way back to the pond at Lighthouse Marsh, now huddling in the bayberry, riding the branches, rocking and swaying as though they were navigating a dory in rough waters, not a shrub in the breeze. Overhead, a half dozen black ducks, and more than twenty teal, wing toward the waters of Big Station Pond, spreading like liquid mercury near the point and signaling what lies ahead: the long flight, the slate and open sea.

Every day the barn and tree swallows crowd the rails that rim the lighthouse tower, the once-thick bands of metal now rusted thin as wire in spots from the salt and spray. First at noon, a solitary swallow sweeps in the unscreened window of the cottage, flies through this inner world like a spirit, and disappears, outside, away. The next day, the visitants-on-wind come later, near nightfall, a pair hovering in the open threshold—Januslike, starting, stopping, the faces of the soul in migration—one venturing boldly in, the other retreating by the same avenue of air that brought it here. Something alien, mysterious—a bird out of place, indoors not out—announces the night, then vanishes, as if pulled by the same curtain of darkness that covers the day.

As summer erodes from the island, taking the longer light with it, an old, undeniable restlessness trembles in the moors and dunes. The garter snakes sun along the trail to the Sound as if to absorb all the heat they can before hibernating, and toads thump in the sandy corridors between the lighthouse and the ocean beach. Even the mice in the keeper's cottage patter about more nervously at night. Though a certain quiet bathes South Monomoy, the grasshoppers tone on and on.

If you didn't know it from the calendar, you could heed it in this sentinel chorus. Summer, like the birds, is pressing on. Fall calms and slows the sandscape, and everything winds down, conserving energy for the cold season ahead.

The emptiness fills me. Everything we normally think of as the passage of time transforms now into endlessness. The sandscape is never the same, changing as though eternity, not erosion, were the force at work here. I can go back and forth across the island—as the clammers and scratchers do from the mainland to the flats and shallows—and still the moving sands and the lush, monotonous beach grass can deceive me, convincing me that I am at the same, exact spot on which I might have been found on some other yesterday afternoon. And then, as I begin again to walk from foreshore to back shore, from drift line to mudflat, the landscape opens like frontier I have never before conquered.

I am not an annual migrant who knows my way. If I want to return to these islands of solitude and contemplation, I know it is in my interest to leave behind a human landmark, a shred of old bandana tied to a bayberry bush, a tatter of towel wrapped around a marker shielding nesting sites, a broken lobster trap half buried in sand. By such fragments I relocate myself in a place I have come to know as intimately as my own breath, and I listen for the serenity that beckons me to return.

On the mainland, blackbirds are staging by the hundreds each evening, settling into the trees just before dark and leaving by morning. From Monomoy I watch the shore slip away, while across the Sound the blackbirds guard the canopy. Every now and then, as I go about the business of island days, the Canada geese drag the V shape of their formation over the sky above me, and to acknowledge our connection and the achievement of their flight, I call out a ritual farewell: Good-bye, good-bye, safe travel.

MIGRATION, particularly the seasonal movement of swarms of birds all over the planet, has been one of the enduring mysteries of the natural world, at least for human beings. The birds, caught up in it, are driven along by choices and decisions that escape us, by knowledge that evolution has made instinct. That, and countless generations of habit, are more than enough to keep many, though not all, birds on the move.

Monomoy is one of the places where tens of thousands of fall-migrating birds stop for rest along a distinct flyway leading south, as far as the extreme southern United States, and, more often, Central and South America. Exactly why they select this land's-end site for sanctuary and a last large feeding becomes clearer once one begins to understand the dynamics and necessity of migration in the lives of great numbers of birds.

Now, at the opening of a new millennium, we know a good deal more about migration than we did for centuries, details about when various species of birds begin their fall flights south and their spring flights back north; what part migration, with all its demands and risks, plays in the struggle for survival for individuals and species; how birds—even the immature making their first flight, solo, separated from parents—know when to go and by what routes. We even know where they go, the far-flung destinations in which they huddle for the winter, if they are able to complete the journey. But it is some measure of the relative infancy of ornithological understanding compared to other life sciences that such information as where migrating neotropical birds end up after their long flights has been known with anything approaching scientific certainty for only less than twenty-five years. And as one biologist pointed out, it was the late 1950s—even as Monomoy was separating from the mainland—before a real outpouring of lucid ornithological work on migration was accomplished. By the time the recorded observations were widely shared and published, it was 1980—more than a decade after humans had walked on the moon and just about the same time that Monomoy was severed into north and south islands by the sea.

Long, long ago, if we had known that we would someday stride on the moon, we well might have expected to run into migrant birds there—so limited, and by today's standards, outlandish, were the theories about how migration worked and where the birds went. Primitive humans, who lived *in* nature to an extent that we would find incomprehensible and unbearable, were so struck by the phenomenon of migration that they imputed it to the gods. In ancient times, an awareness of birds' seasonal movements and the north-south patterns of flight was widespread enough for Old Testament writers to refer to migration as evidence of divine action and constancy (Job 39:26) as well as human's proper relation to the creation and its creator (Jeremiah 8:7). Other early, and poetic, observations show up in Homer's *Iliad* as well as in the works of Anacreon, Hesiod, and Aristophanes.

But it was Aristotle who offered the first so-called scientific explanation of migration, recording observations of various species' movements from north to south as well as from higher to lower altitudes—all presumably undertaken by birds searching for warmer weather and greater stores of food. But Aristotle also held that not all birds migrated; many, he thought, hibernated instead, taking refuge in holes and burying themselves in marsh mud until the return of warmer weather. However anachronistic his theory might seem to modern ornithologists and backyard birders, it prevailed in one form or another for centuries, its errors compounded by later scholars.

Fascination with the seasonal flights of great numbers of birds—particularly cranes, swallows, storks, pelicans, and turtledoves—persisted, often uncritically recorded in natural history works into the Middle Ages. But one of the most charming biological records, to assign that designation loosely, came from a 1555 French translation of the work of Olaus Magnus, the Archbishop of Uppsala, who wrote some intriguing—if fanciful—impressions about migrating swallows:

> Several authors who have written at length about the inestimable facts of nature have described how swallows often fly from one country to another, traveling to a warm climate for the winter months; but they have not mentioned the denizens of northern regions which are often pulled from the water in a large ball. They cling beak to beak, wing to wing, foot to foot, having bound themselves together in the first days of autumn in order to hide amid canes

> and reeds. It has been observed that when spring comes they return joyously to their old nests or build new ones, according to the dictates of nature. Occasionally young fisherman, unfamiliar with these birds, will bring up a large ball and carry it to a stove, where heat dissolves it into swallows. They fly, but only briefly, since they were separated forcibly rather than of their own volition. Old fishermen, who are wiser, put these balls back into the water whenever they find them.

More reliable natural history observations would not come until much later, when naturalist-ornithologists took the first steps toward debunking the long-held theories about hibernating birds. But intensive investigations into bird migration were delayed for centuries because of scholarly preoccupations with what were believed to be more crucial flights, such as that of the human soul from this life to the next. Careful, accurate observations of birds in the field and under then-rugged laboratory conditions did not come until the eighteenth century, and it was not until well into the nineteenth century that the theories of winter torpor and hibernation of certain species of birds—principally, swallows—were refuted. The real flurry of sound ornithological writing on the subject began to show up only in the mid-1900s. In fact, much of the most solid data and interpretations—particularly those related to North American migratory birds—surfaced as late as the 1950s and 1960s, aided by findings derived from the practice of "ringing," or banding, birds with aluminum rings; the use of radar; and the observations of species such as pigeons in homing experiments.

I THOUGHT A LOT about migration while I made my way across the island and the years—how the creatures seem to accept change viscerally, in a way I usually cannot. Birds weave their patterns of travel along avenues of air that can carry them a few hundred miles south or halfway around the globe, depending on the species. From down below, here on the limb-heavy ground, the principal flight is one of fancy, and I take it, imagining the miracle of a tiny bird, for which inner prompting and outer action are inseparable. This unassailable motivation to travel a thousand miles in spring and fall in response to an internal imperative represents, in birds, a seamless bond of inner being, outer expression, and encompassing nature. It is a harmony humans might strive for, if only we could silence our conscious, hungry minds enough to follow something more animal, something more avian, in our blood memory.

Failing that, we watch from an inferior vantage and try to guess at what is going on. The questions—because they often disclose as much about us as the subjects of our inquiry—teach us, even if with a different instruction, as their answers do. "Why move?" we ask, forgetting for a moment that we ourselves are the edgiest inhabitants on the planet, lurching from one place to another and subjugating too much along the way, without the excuse of the animals, that our behavior is designed to promote survival. Our movements, encoded in greed and aggression, threaten to collapse the very environments that sustain us, while overhead, without even trying, the birds show us how to soar.

Through more watchful, less anthropocentric, attention, we have become aware that many creatures follow more or less predictable migratory paths, according to rhythms and timing that evolved into more-or-less constant patterns over hundreds of thousands of years. Squid, for example, migrate along the coast of Europe. Whales, seals, sea turtles, and salmon accomplish pilgrimages over hundreds and thousands of miles in oceans or up rivers in various parts of the world. And despite their fragile weight, insects—especially moths, butterflies, dragonflies, and spiders—turn out to be astonishingly strong, adept, and successful travelers.

Still, it seems miraculous that a tiny bird, such as a swallow weighing a few ounces, can navigate thousands of miles and negotiate winds, storms, and other hardships of weather and long-distance travel. We continue to ponder just how birds—particularly immature young that have never made the flight—know when to go and by what route to travel, from, say, northern Canada to South America. Since so many fly at night, by what compass do they persist on course? What specialized anatomy and physiology enhance the birds' chances for survival? And why, if weather is favorable and food abundant, would a bird choose to face harsh conditions and migrate?

There still is a lot we do not know.

What we have learned through decades of close observation and record keeping is that the general patterns of birds' movements, however revealing, do not apply to all species equally or to all ages of birds in the same way. For example, in some species, immature birds might migrate south while adults remain farther north. Or, birds that occupy the southern edge of a range, where winter weather is less taxing, might remain in those grounds rather than move. Some birds make their long-distance flights in bursts of travel, stopping to feed and rest along the way, while others fly nonstop. The blackpoll warbler in autumn bulks up along the Massachusetts coast, putting on fat weight for a 1,900-mile trip. Likewise, terns are not only among the longest distance fliers, in some cases, circumscribing the globe; they are also in the air for very long stretches. Because their routes carry them over water, they must stay aloft most of the time, but if too stressed, they can rest on the ocean's surface.

Broadly speaking, birds migrate north to south; but there are east-to-west migratory paths as well. Some fly by day, others by night. Scientists now know that the flights are prompted in part by birds' "internal clocks," one of which is timed to daily rhythms and the other to annual rhythms. A bird's hormonal system is regulated by the combined effect of shorter days (and less light) and its yearly clock. As the time for migration approaches, the bird adds body weight and becomes restless. In the wild, it soon begins to migrate; but if it is caged, it will become agitated, fretting in the part of the cage that faces in the direction in which, if freed, it would be flying.

We rely on other natural phenomena, including the moon, to help us attain the mind of a bird. For more than a century, human observers of nocturnal migrations have used the moon to illuminate navigations obscured otherwise by darkness. When the moon is full during a period of intense migration, it is possible simply to watch the orb through a telescope—as though it were a spotlight—and count birds moving through the area. If the wing beats of the birds are slow enough, it is even likely that experienced birders will be able to identify which species of bird is crossing over the path of moonlight at a given moment.

But the birds probably rely on different sextants.

After nearly 150 years of study, ornithologists know that birds migrating at night often use the stars to direct their course. For a long time, the very notion that birds might be able to navigate using the sun, the stars, and the Earth's magnetic fields seemed preposterous—so certain were we that human intelligence outstripped the "lower" creatures. But years of testing and tracking have taught us something we hadn't predicted: namely, that birds have capabilities with which our own, self-revered faculties, operating independent of technology or machines, cannot compare.

We have run species of birds through all kinds of scientific hoops, putting indigo buntings in a planetarium with various projections of the sky overhead to trace the ease with which they were able to pinpoint the North Star. We have placed robins in cages and used our powers of reason and inventiveness to create conditions under which magnetic coils artificially induced a sense of the planet's magnetic field, all to find out if, and how quickly, they would reorient their flight using the changing field. We have set migrants in cages with mirrors that would "bend" the direction of sunlight entering the birds' immediate sphere, and we have watched as they adjusted their flight to scientific sleight of hand.

I do not mean to denigrate scientific investigation or the attempt to isolate answers to puzzling questions that cannot help intriguing us, even if we are only half awake to the world around us. We share the planet with the birds—and they are among the first "others" to tease us out of ourselves long enough to appreciate the natural world, which cradles us and over which we hold such awesome power of devastation. While we take pride in our ability to walk upright on our hind legs and our power to make machinery hasten over highways and continents, the birds ascend to the skies, following their own routes, charted by internal clocks and maps that it has taken us millennia to study.

And in their innocent, instinct-imprinted travel, they move more than themselves.

The sight of birds flying south in fall or returning north in spring touches an emotional timing deep in us, a peal of deep, shared, planetary rituals and profound individual hope. The wing—though it seems an appendage alien to ours—still announces a kinship we understand. A feather, for all its difference from the fur nearer our own, is yet able to rouse a flutter in the earthbound imagination and the petrified human heart.

Feathers—molted or lost in fight or flight—are one of the first things that spark humans' interest in birds. Finding a primary from the wing of a jay or a crow's tail feather can be the introduction to a lifetime fascination with birds and, perhaps later, with other animals. I remember, as a child, happening on an iridescent purple feather as I shuffled through leaf-littered gutters along the urban streets leading home from school. Over time, I collected specimens from doves, sparrows, finches, robins, jays, cardinals, starlings, and even a mallard that nested by the polluted river that ran through a nearby forest preserve. I was hardly out in the wilds, but even in a metropolis, you are in the arms of nature—whether you know it or not. The birds led me to look at nature as something more than scenery or a pretty surrounding to break up commercial development. Because of the creatures flying in and out of my days, I began to see all of the natural world as alive, inspired with a life force I shared but did not control. I came to witness how each of us has a place and part—a connection, some relation, but no absolutely superior position in creation.

A robin, almost ready to fledge, first communicated that bond to me. The plump little bird had become separated from its mother and its nest when I found it one early summer morning, peeping frantically from its hiding place in the lily of the valley at the side of the house. For what seems, in memory, like hours, I sat in the sunlight on the sidewalk, in a T-shirt and rolled-up overalls, my bare feet poked into the dirt of the flower garden, watching and listening to the bird, hoping the mother would return. I did not realize that my very presence—small and harmless though it might have seemed to me—would have been danger enough to keep the adult bird away.

Finally, I gave up waiting and got out the predictable rescue equipment of childhood: a shoe box lined with ripped newsprint, tufts of grass roughly fashioned into a cup nest, a bottle lid filled with water, and an eyedropper. I even went digging with a trowel for some earthworms, intent on doing everything I could to try to ensure the bird's survival.

Later that morning, my family's live-in housekeeper—an older woman who had grown up on a farm in Michigan—lifted the bird out of the flowers and set it, trembling and silent, into the fabricated environment of safety. Until early evening I kept it, cradling the box in my arms, toting it here and there in the yard, talking to the "infant," making the rounds, but not touching the bird's intricate feathers. At dusk we decided to put the bird back where we'd found it in hopes that the mother might still return; and that night, I fell asleep to the sound of peeping. In the morning the bird was gone.

Even today, I can conjure up that robin—the first I ever saw close at hand—as though I had only just finished studying its subtle colors, its grave, blinking eyes and gaping beak. Psychology gives me a guess about why the fate of the bird seemed to me, then, so like my own, but the sad insight fades under the brighter gleam of a child's innocent curiosity and one high human instinct—to help, not hurt. In the harsh light of living, anything can become a lasting trait; thank God for the stirrings of marvel that endure, over time, as deep wonder, respect, and grace.

It can all begin with something as simple as a feather, the strewn attire of a fellow traveler, evolved over countless thousands of flights to handle the battering of high winds, long-distance journeys, hard rains, and bitter cold, perfected so that a frail bird—a mystery large enough to fill the palm of the human hand—could lift itself over trees and the edifices of men, carrying on its light and powerful wings the imagination and hope of a child.

THUS, AS THE SKY fills with wings, fall approaches. At this palpable threshold between summer and fall, the songbirds will feed for the day, and, if exhausted from flight, perhaps longer; the shorebirds will linger for a couple of weeks at most, then continue on their various courses to various destinations, including Mexico, the West Indies, Central and South America. And here on one far shore of New England, on the last border of summer, paradise opens for birders.

Mid-August to mid-September is the peak period to catch sight of migrating birds on Monomoy. Ornithologists deem it the ideal time during which to glimpse both regular migrating bird species and even the so-called vagrants—birds that turn up in unexpected places, blown off their usual flight routes. But fall migration in the will and wings of the birds actually begins much earlier, as early as late June, when the first shorebirds arrive on Monomoy from sub-Arctic Canada. The timing is so precise that a person can pretty much clock the days by the birds, rather than vice versa. It is reasonable to assume, for example, that the Hudsonian godwit will complete the Hudson Bay to Cape Cod leg of its flight on or about the Fourth of July by the human calendar. From then until well into November, great waves of birds will swell over the island, some staying slightly longer than others. Birds of prey and sea ducks are among the last to reach this island stopover point. But on a good day during the height of fall migration, alert birders, making their own course across the Sound to the islands, can glimpse upwards of a hundred species.

Even so, birding today on Monomoy is not the event it used to be. Blair Nikula of Chatham, a self-taught ornithologist and past president of the Cape Cod Bird Club, has been birding for decades on Monomoy, and he senses the difference acutely. He made his first visit to the islands on an Audubon Society birding tour more than thirty years ago, when reports of 100 to 125 species in a single day were common among experienced birders. "It used to be you'd see hundreds of warblers," he says. "Now, if you see dozens . . ."

The explanation for the decline in numbers—particularly of migratory songbirds—is not thoroughly understood, but one clear reason is loss of habitat in the birds' summer and winter territories. How much habitat has been lost in North America and how much in the tropics and the Southern Hemisphere are matters of debate, but there is no arguing about the effect on birds. Even so, Monomoy is considered one of the best birding sites in the Northeast—especially during fall migration—and the islands still draw birders from all over the country. Nikula himself makes as many as eighty migrations a year from Harwich, on Cape Cod, over to Monomoy for bird-watching. He is drawn, of course, by the numbers and variety of birds; the birds, on the other hand, keep showing up in part because the islands have such a variety of habitats, including freshwater ponds and marshes, tidal flats, moors, dunes, and thickets of low brush.

Monomoy's importance for migrating birds has been recognized for decades. Some of the patterns of migration—and the predictability of their dates of occurrence during spring and fall—have been documented since the first quarter of the century, notably through the work of Massachusetts ornithologist Ludlow Griscom and, later, by his associate Dr. Norman P. Hill. The physician, who had accompanied Griscom on numerous Monomoy birding excursions, compiled material from Griscom's unpublished journals and field notebooks as well as his own scientific observations and data to produce *The Birds of Cape Cod, Massachusetts.* To this day, it is considered an authoritative regional guide for birders. Toward the century's end, Massachusetts Audubon ornithologist Wayne Petersen and biologist and ornithologist Richard Veit updated Hill's earlier work in the comprehensive edition of *Birds of Massachusetts* In such collections are written the histories and legacies of nations other than human, in language we can translate, interpret, and, in time, begin to understand.

Roger Tory Peterson, the artist who transformed the world of birds and the humans who watched them by arming himself with paint, ink, pencils, paper, brushes, and binoculars—instead of the authority of a collector's gun—accompanied Griscom on numerous trips to Monomoy. In his introduction to *The Birds of Cape Cod, Massachusetts,* Peterson recalled the heyday of Cape Cod ornithological finds in the 1930s and 1940s—a period still considered by many to have been the golden era of Cape Cod birding. Griscom, who is remembered for his skill in sighting and identifying birds strictly by field marks, hosted innumerable excursions along the Outer Cape, embarking from Harvard Square in Cambridge or, alternatively, from his house at Sears Point, Chatham, which was used as a base from which to head to Monomoy or Nauset Beach and Marsh. There, groups of professional and amateur ornithologists would spend the day scoping predictable species and documenting exceptional sightings.

"A high point in any Cape Cod field day with Griscom was a drive down Monomoy," Peterson recalled. "He was phenomenal in spotting rare sea birds beyond the breakers and vagrant waders on the mud flats. His conversation was salty and original, punctuated with such stock comments as: 'Let's stop here and flap our ears. . . . Let's get out of here, it's getting dull . . . just dribs and drabs left. . . . Check me on that one. . . . What's the tide schedule? . . . Now somebody find a bird with some zip in it. . . . That's a weed bird. . . . Please lower your voice to a howl. . . . I don't like the look of that bird. . . . Unprecedented. . . . Put it down to sheer ignorance, incompetence and inexperience. . . . We got skunked on that one. That's a 10-cent bird. . . . Well, we didn't do so badly. . . . Having a good time?' . . . etc."

It is sociability as well as science that has fueled the drive for extraordinary birding experiences along the Cape, up to and including the present-day work of naturalists and wildlife biologists from the Massachusetts Audubon Society's Wellfleet Bay Wildlife Sanctuary, educators from the Cape Cod Museum of Natural History, in Brewster, and the Cape Cod National Seashore. Avid birders congregate regularly under the auspices of the Cape Cod Bird Club and private natural history enterprises up and down the peninsula.

Ornithologists' and birders' fascination with Monomoy was especially intense in the 1960s, when the Massachusetts Audubon Society, under a cooperative agreement with the Fish and Wildlife Service, led beach-buggy tours of the island. Wayne Petersen, then in college, served as one of the guides from 1966 to 1968 and remembers the era as a time that held, he says, "some of the fondest birding memories of my life." Tours still run regularly, made possible by the boats that go out to the north and south islands at different times of the birds' year. Dune buggies, of course, are gone. In accordance with the islands' wilderness-refuge status, travel on the islands is now done on foot.

In contrast to the ever-changing human relationship to Monomoy and the creatures who establish themselves there permanently or for a brief rest when migrating, the routes of the birds' migration remain relatively constant. It was Griscom and Hill who described the "flyways," or lanes, that migrating birds used when crossing the landmass of Cape Cod. They hastened to point out that these designated flyways hold true generally, but not exclusively, and that birds move over a broad front when making their seasonal travels. The flyways humans have pinpointed represent not hard-and-fast rules, but rather patterns and concentrations of birds.

One north-south flyway described by Hill in *The Birds of Cape Cod, Massachusetts* runs along the inner coast of Cape Cod, spanning nearby land and water. It covers the western edge of Cape Cod Bay, crossing the peninsula at the Cape Cod Canal and encompassing all of Buzzards Bay. The second major flyway, the Outer Coast route, extends south from the Outer Cape at Provincetown all along the National Seashore and Monomoy and west to Brewster and Harwich. There is a greater concentration of birds along this second flyway, but greater numbers move on the inner-coast route. A third route crosses over the Cape in the area of Barnstable and Hyannis and picks up the spillover of birds from the first two, more commonly used, flyways. A fourth route is across open sea and runs along a north-south or northeast-southwest direction east of Cape Cod. Its closest points of contact with the Cape extend from Eastham to Monomoy.

Since so many birds fly over or near Monomoy—and because weather can effectively modify the routes and cluster birds in concentrated fallouts along the coast—birders who make the voyage to the refuge are seldom disappointed, particularly during fall migration. It is estimated that twice the number of migrating species, and greater overall concentrations, move across Monomoy in fall as compared to spring.

During August and September it is common to witness birds by the hundreds, and sometimes, thousands, congregating in freshwater ponds and tidal flats or marshes on the islands. South Monomoy is the preferred site for birding in autumn.

Four principal groups of birds can be seen on Monomoy during fall migration.

Shorebirds are the earliest migrants in evidence on the islands, and they arrive in two great waves, with a notable break in between. The initial surge of migrants crests in late July and early August; the second, in late August and September. The first wave tends to bring adult birds migrating ahead of their young, while the second is characterized by a larger number of immature, juvenile birds. From late July on, shorebirds spotted on Monomoy include ruddy turnstones, black-bellied and semipalmated plovers, short-billed dowitchers, and red knots. One dramatic exception to separate adult and juvenile migrations is the dunlin, which stays in the Arctic longer than most shorebirds seen in that region. Ultimately, both parents and young take flight together, almost simultaneously. Some of the uncommon shorebirds that show up on Monomoy's shores in autumn—frequently around Labor Day—include Baird's, buff-breasted, and western sandpipers; long-billed dowitchers, and American golden plovers.

For migrating songbirds, weather and winds play an important role in determining how many will end up stopping on Monomoy in the fall. Because of a phenomenon known as "wind drift," in which tailwinds from the northwest cause immature birds to veer off the usual, preferred course over land, what Wayne Petersen describes as "fallouts of songbirds" can be seen for brief periods, particularly on South Island. These fallouts occur when a cold front of dry, arctic air from the northwest sweeps birds east, suddenly flooding Monomoy with flycatchers, vireos, swallows, thrushes, many species of warblers, grosbeaks, and northern orioles.

Many of these species are nocturnal migrants and will fly back north to Morris Island after dawn to reorient themselves and get back on the mainland, over which their usual flyways extend. Consequently, the best time to see these birds is at dawn or shortly after. More than 90 percent of the songbirds that show up under these conditions are immature birds, and their presence has more to do with inexperience and navigational error than habit or desire. And while birders are heartened by the possibility of a prevailing northwest wind the night before a trip to Monomoy, it is by no means ideal for the birds themselves. They drift off course and have to retrace their flight path to get back on track—a lot to ask of a small, usually immature, bird. Many perish. Chatham might be prime real estate for people who come to Cape Cod to visit or live, but it's not the place to be if you're a young bird on your way to Costa Rica.

Among waterfowl, both adult and juvenile American black ducks, blue- and green-winged teals, mallards, gadwalls, and northern shovelers are evident on and around Monomoy by late August and early September. A steady influx of teals, northern pintails, gadwalls, and American wigeons occurs through September; and October and early November mark the peak for eiders, oldsquaws, and scoters.

Birds of prey—including northern harriers (some, no doubt, raised on Monomoy or nearby), sharp-shinned hawks, merlins, and peregrine falcons—are among the more majestic birds that can be seen throughout the fall migration. Virtually all these birds will quit Monomoy and fly farther south. But some birds do stay, for better or for worse, all year-round: gulls.

NO BIRDS HAVE BROUGHT more controversy to Monomoy than gulls. No species has been more reviled for its aggressive, territorial displays—except, possibly, the humans who have tried to curb the advance of the large gulls over habitat other birds require for nesting and raising young. On Monomoy, the fates of all the birds are inextricably linked—both among species and to our own.

The ascendancy of great black-backed and herring gulls in the last half century has marked a kind of avian Manifest Destiny, spurred, ironically, by the birds' uncanny ability as hangers-on to humans. Gull populations on Monomoy, for example, had an imperialistic level of success because for many years, the birds had a seemingly inexhaustible supply of food at open landfills stocked with human garbage on Cape Cod and near the East Coast. Even after landfills on the Cape had been capped, the birds—especially those migrating south for the winter—were better able to survive the cold months because they forage in the discards from fishing fleets, restaurant garbage bins, and roadside litter. The stronger and more belligerent gulls extended their range and population, overwhelming Monomoy and spelling decline for the endangered piping plovers and common and roseate terns. People and pets participated directly, too, in the undoing of established nesting areas of plovers and terns on the Cape. But those trends have begun to be reversed—though only after an intense and bitter public debate.

The situation for the endangered and threatened birds on Monomoy is, in every way, dramatic, and at times sensational. On an island where there is not much room, anyway, nesting habitats for shorebirds and waterbirds are at a premium. The birds need beach and dunes, where they nest in scrapes, small, wing-brushed areas in the sand. Piping plovers and terns, which are shy and skittish birds even under the best circumstances, can be displaced quickly and easily from their nests by inadvertent disruption or actual threat. Humans, coyotes, gulls, and black-crowned night herons all interrupt the nesting cycle of terns and plovers and can cause territories to be abandoned.

Great black-backed and herring gulls, compared to any number of other birds on Monomoy, are massive birds, capable of fierce territorial battles and an almost frightening penchant for survival. They not only crowd other birds and colonies off landscape they claim for breeding and nesting, but are also given to feeding on smaller birds, including the young of other gulls. If you consider only the gulls' length from bill tip to tail tip and then compare it with the sizes of species vying for nesting sites or living space on the refuge's beaches, the gulls' advantage is immediately obvious: great black-backed gulls measure from 28 to 31 inches, followed by black-crowned night herons at 23 to 28 inches, and herring gulls at 23 to 26 inches. In contrast, terns average between 13 and 17 inches (the roseate and arctic at 14 to 17 inches, the common at 13 to 16 inches), and plovers run between about 6 and 13 inches (the black-bellied plover commonly reaches 10½ to 13½ inches; the piping and semipalmated measure from 6 to 7½ inches). Immature gulls and herons, once fledged, exceed the size of adults of all these other species, making it very difficult for terns and plovers to contest territory, habitat, or food. Terns, which are skilled fliers and divers—and, coincidentally, some of the most heartening birds to observe because of their precision, beauty and grace—are also anxious and easily alarmed, regardless of whether the stalker is a gull, heron, coyote, or human. And plovers can be downright reclusive; with their light coloring and small size, they appear to the untrained human observer to be little, irregular eruptions of the beach as they scurry about in fits and starts.

A look at the broader picture of what has been happening for a decade or longer, not only on Monomoy but also all around it, yields a clearer idea of what the birds are battling. The place to start, perhaps, is with raw numbers—population statistics—moving from counts of humans to tallies of bird species:

In the twenty years between 1960 and 1980, the year-round population of Cape Cod doubled from 70,000 to nearly 150,000. And by 1998—the most recent year for which the U.S. Census Bureau has available statistics—that number jumped again by nearly 40 percent, to 208,000, the equivalent of adding to the peninsula a town the size of bustling Yarmouth in the mid-Cape. Summer is even more crowded. An estimated 300,000 to 400,000 summer residents and visitors flock to Cape Cod from June to September.

While residents and public officials on Cape Cod tried to accommodate an exploding population and the land development that comes with it, the gulls followed their usual opportunistic patterns of taking advantage of a new food supply—humans' garbage. The discarded food of hundreds of thousands of people represented, at open landfills an easy flight away from Monomoy, a nutritious smorgasbord for omnivorous gull populations, which regularly include carrion and refuse in their diets. Unknown in New England in the nineteenth century, herring and black-backed gulls had by 1920 only just begun nesting on Monomoy. By 1950 they were commonplace, and by the 1980s they had overwhelmed the refuge.

By contrast, the count of stressed birds in the early 1990s posed a staggering juxtaposition—roseate terns, zero; common terns, zero; piping plovers, three nests, eight fledglings. Until 1995, the numbers of piping plovers and terns on Monomoy showed little improvement; gulls, biologists believe, were the culprits, usurping plover nesting sites and scattering tern colonies. When the Fish and Wildlife Service introduced vigorous management efforts, including a gull poisoning program, in the mid-to-late 1990s, the tern and plover populations began rising.

But there were other offenders—namely, coyotes. No one knows exactly how many coyotes have entrenched themselves on Cape Cod in recent years, but even anecdotal reports indicate that the numbers are increasing and the canid's range is extending. They have been spotted in every town on mainland Cape Cod and have been known to take small household pets as prey. And in 1998, the very thing that naturalists had said was unimaginable happened—a child was attacked by a coyote on Cape Cod. The Cape's coyote population has dispersed, reaching Monomoy, where the animals have attempted to den. There, food—ordinarily small rodents and fledgling birds—is plentiful, and open habitat is uncontested by other, mostly carnivorous mammals.

Descriptions of terns' and plovers' preferred habitats for nesting read like a prospectus to attract not shorebirds, but tourists. Piping plovers, for example, nest along the Atlantic on coastal beaches above the high tide line, on sand flats at the ends of sand spits, and on barrier islands. The birds seek out gently sloping fore dunes, blowout areas behind primary dunes, and washover areas cut into or between dunes—the very places people haunt from late March to mid-August, when birds occupy these areas for breeding, nesting, and rearing of fledglings. Flat, overwash areas, such as the Powder Hole on South Monomoy, are ideal sites. And while competition among the birds for food and nesting space would be conflict enough, human intrusion into their habitat is one thing terns and plovers cannot tolerate. Without help, these birds on Monomoy "have got a hell of a tough row to hoe to survive," according to the former refuge manager Ed Moses.

To bolster the chances of survival and nesting among plovers and terns, the Fish and Wildlife Service closes the beaches on which they nest from April 1 to the end of August. Boaters are turned away by refuge officials, and people who repeatedly violate the closings or cross into posted areas of the islands can be ticketed. Officials acknowledge that trying to convince the public to make room for birds along some of the most desirable beaches in New England is a very slow, uphill battle. To encourage understanding, the Fish and Wildlife Service has tried to promote educational law enforcement over sterner measures. More strict enforcement is undertaken only when people are persistently recalcitrant or militantly indifferent to the birds and the measures to protect them.

In recent years, public disregard for posted areas and the use of Monomoy for overnight camping has prompted more vigorous patrolling by refuge officials and more citations for violations. Illegal camping traditionally has been a problem on the island, but Fish and Wildlife officials are reluctant to respond by issuing tickets or fining campers because such measures tend to further alienate people the service would like to enlighten about the birds and their island habitats. "What do we gain but a lot of hatred and resentment?" Moses asked rhetorically during an interview at Monomoy Light in the mid-1990s, before any widespread culling of gulls had occurred. Through education, he hoped, people might come to understand that "terns and plovers have as much right as you do to be on this earth."

But it was going to take more than limiting human activity on Monomoy to return the island refuge to a full range of seabirds, shorebirds, and songbirds. Under federal law and according to the articulated goals of the National Wildlife Refuge System, refuges like Monomoy are required to "preserve, restore and enhance in their natural ecosystem (when practicable) all species of animals and plants that are endangered or threatened with becoming endangered." Federal laws and recovery plans for the imperiled birds compelled wildlife officials on Monomoy to intervene to help bird species being forced out by gulls.

Fish and Wildlife officials knew that fulfilling the mission would be difficult in a public climate more likely to favor protection of all birds and animals than elimination of targeted species. But generating sympathy for threatened and endangered birds didn't seem to be an impossible public-information goal in the fall of 1993, and if public sentiment could be mobilized for terns and plovers, the less palatable methods of poisoning and removing gulls from part of the island might be successfully argued. But Fish and Wildlife officials, faced with the opposition of animal-rights advocates who wanted to quell any human intervention in the wildlife sanctuary and people who wanted greater freedom to use refuge beaches, were soon to find out just how difficult and controversial their federal mandate for management and species diversity could become. In 1996, a three-year gull-control program, barely under way had whipped up so much public furor that it had to be curtailed—despite its ultimate success in realizing the projects' goals: namely, to return endangered and threatened birds to the islands.

"Species diversity," in this instance, specifically addressed the plight of terns, piping plovers, black skimmers, and laughing gulls, and "management" meant poisoning and shooting great black-backed and herring gulls as well as destroying eggs on nests and harassing the birds so that they would have to abandon their nesting sites. The methods worked. In two years, slightly more than 1,700 pairs of nesting herring and great black-backed gulls were eliminated from about seventy-five acres on South Monomoy. Common terns, which had been so stressed by gull populations that they were almost wiped out, rebounded from an estimated 50 or so pairs in 1995 to more than 650 pairs in 1997.

But the terns were just getting started. In 1998, their numbers jumped to more than 2,350 nesting pairs, and in 1999, skyrocketed to nearly 5,500 nesting pairs. Additionally, the terns almost immediately moved in on interior dune nesting sites that long had been monopolized by the larger gulls.

"The remarkable thing is how fast they responded—it's thrilling," says Anne Hecht, an endangered-species biologist with the U.S. Fish and Wildlife Service in Sudbury. The fundamental problem for the terns attempting to nest elsewhere had been that once pressured by a predator—gulls, coyotes, or humans—they had nowhere to flee. Even a small portion of cleared island habitat made, literally, all the difference in the world for them. The rebounding tern populations, biologists agree, sent a clear message about how desperate the birds were for places to nest.

Now, with enough space to breed and nest, the terns are likely to maintain or even increase their numbers. Monomoy is an ideal place for them because the waters around the island provide a rich supply of sand lance. Dazzling aerialists, terns frequently hover in midair and dive headfirst to take these small fish that are the staple of their diet. And now, as a new century opens, the sight of their acrobatics in the air has become common once more.

The success of nesting piping plovers has been less spectacular and is still tenuous, even on Monomoy. By 1999, more than twenty-five nests were counted, but biologists who have long studied the species' status along the coastal United States believe that the island refuge has a substantial unrealized potential to host many more.

Beyond Monomoy, where the jury is still out on piping plover recovery, numbers of the birds nesting in New England are beginning to level off, and wildlife biologists are not entirely sure why. The impact of gulls on declining piping plover populations and nesting areas has been documented on Block Island, Rhode Island; South Metompkin Island, in Virginia; and Long Point, Ontario. But while biologists and ornithologists ponder the fate of piping plovers and try to enhance the species' chances for survival, they are tallying unexpected boons to other birds brought about by gull-control measures. Even songbirds have responded to the change in the habitat. Savannah sparrows and horned larks—which had not been seen in 1996 in the area of South Monomoy covered by the management program—showed up the next summer. And by 1999, roseate terns, black skimmers, and laughing gulls—smaller and less bellicose than other gull species—were occupying acreage and sites in which they had rarely been seen in recent years.

Despite these gains in endangered and threatened species, animal rights activists and many birders were outraged by the gull-control program, and they spoke out emphatically—and sometimes violently—against the gull poisoning at public hearings and in private contacts with the Fish and Wildlife Service. During the summer of 1996, the first year the plan had been implemented, the program erupted into chaos. For about ten days, poisoned gulls were flying from South Monomoy to freshwater ponds in Chatham to die—a process that sometimes took hours—with the public there to watch. In response, members of the U.S. Humane Society arrived in Chatham to help euthanize slowly dying gulls. Reporters from local, regional, and national news outlets were there to chronicle—and further publicize—what was already a painful situation from every point of view. Some people who opposed the poisoning became so angry and alienated that they began warning of violence against Fish and Wildlife officials; death threats were not uncommon. Even schoolchildren were protesting the killing of the birds, writing to newspapers and calling radio and television news programs to air their grievances.

Stephanie Koch, a wildlife biologist who completed her bachelor of science degree at the University of Massachusetts in Amherst in the late spring of 1996, went to work for the Fish and Wildlife Service at the Monomoy refuge that turbulent summer. She remembers it as the "hardest thing" she ever had to do; and what she is referring to is not only being vilified by the public but also the sheer fact of having to destroy any birds at all. Believing that the program was the only workable approach to maintaining diversity among bird populations didn't take away the sorrow that came with killing gulls—or later, in 1998 and 1999, with the killing of coyotes.

"A lot of people are upset with lethal control of predators," she says. "We're living in a world that has to be managed. The whole concept of 'nature' isn't there anymore." The overwhelming advantage of certain predators—whether gulls or coyotes—over other species represents "maybe . . . something we can fix." The goal really is to have safe nesting sites for the greatest number of bird species, she says. The success of that will have to be judged in the long term.

Bud Oliveira, a U.S. Fish and Wildlife Service veteran who, in 1997, succeeded Ed Moses as refuge manager, faced the public outcry his first week on the job. Because the service is charged with managing refuges while maintaining and enhancing diversity of species, he believes some of the gulls had to be eliminated so that piping plovers and terns could have what he calls "a stress-free habitat" and a real chance to survive, breed, nest, and raise young. While he understands the impulse to "let nature take its course," he finds it too simplistic an answer—and too late—for a place like Monomoy, where human intrusion has been a reality for four hundred years at least, and where wildlife management has been a fact of life for nearly sixty.

"We've changed the world so much since European man came," he says. "We've destroyed habitat, killed off species. The whole balance is out of whack. We've tinkered so much that now we *have* to tinker."

PIPING PLOVERS are the sand-colored wading birds described by Roger Tory Peterson in his field guide to eastern birds as being "pallid as a beach flea or sand crab." The birds are considered by the federal government to be a threatened species, a designation indicating that, without human intervention, the birds would likely become endangered in all or part of their range. Endangered species, in turn, are those that, unaided and unprotected, would face extinction. To help piping plovers along the Eastern Seaboard, a large network of government agencies, private groups, and nonprofit organizations have joined together to protect and improve habitat, reduce native and introduced predators, and manage harassment from humans and their pets.

Since 1988, the Fish and Wildlife Service biologist Anne Hecht has coordinated the U.S. Atlantic Coast Piping Plover Recovery Program, a cooperative effort extending along the plovers' breeding range from Newfoundland to South Carolina and into their wintering habitat to the south. As a biologist, she is sensitive to the birds' importance not just in terms of avian diversity but also because, she says, piping plovers are such a "good indicator of an ecosystem that's in a lot of trouble . . . the beach. When we protect piping plovers, we provide benefits to other [stressed] species and the beach. But these aren't things you do once and walk away. If you stop doing them, you'll be right back where you were. . . . The population has been recovering, but the threats are pervasive and unrelenting." Coastal residential and commercial development, oil spills, and humans indifferent to the risks the birds face all have contributed to the continuing dangers plaguing the species.

Like the roseate and common terns on Monomoy, piping plovers need a safe habitat for nesting. While they usually choose sites with sparse vegetation, they occasionally occupy areas in which dense stands of beach grass have taken hold. Many of these birds, especially males, tend to return year after year to the same sites, once their affinity for the location has been established.

In Massachusetts, piping plovers return to their breeding grounds in the third week of March to begin establishing territories and start courtship. Their rituals are elaborate: Males vocalize while bobbing their heads up and down and performing elliptical or figure-eight flights over their territories. They make shallow scrapes in the sand and toss in shell fragments—all to announce their intention to reproduce. By late April and early May, courtship is accomplished and the females lay eggs in nests as insubstantial as churned sand. Chicks begin hatching anytime in late June. Rearing of fledglings continues from mid-June through early July. And if the first breeding is unsuccessful, pairs will breed again. The result is that the birds may be involved in nesting or rearing young through late August—high season for human use of the islands, as well.

Since the inception of the recovery program in the late 1980s, piping plovers have been faring better in Massachusetts, but the total numbers are still quite small. In 1991, 160 pairs of piping plovers were counted in Massachusetts alone; in 1992, 213; in 1993, almost 300 pairs. Five years later, that number reached almost 500 pairs. Even so, the recovery for plovers traditionally has not been as good on Monomoy as in other parts of the state. Most of the plover recovery in Massachusetts has occurred along the Outer Cape, north of Monomoy; on Sandy Neck; in Duxbury, on the South Shore; along the North Shore; and on Nantucket and Martha's Vineyard. Protective measures used in these areas epitomize the recovery strategy. On beaches, symbolic fencing of strung twine warns humans to steer clear of nests; wire fencing is erected around established sites to keep predators out; and the use of motor vehicles is controlled. Nature, too, has cooperated by renewing habitat through storms that cause overwashes and flatten beaches.

From year to year, and from one week to the next during a summer season, the plovers keep trying to nest and establish themselves. Once they have eggs on the ground, they're not likely to abandon their nests voluntarily. If disturbed, however, their natural behavior is to pretend to have a broken wing and hop away, leaving the nest temporarily undefended. When that happens, threats to successful incubation only multiply. Humans unintentionally cause grave problems simply by showing up to enjoy an afternoon at the beach. The average stay by boaters or sunbathers has been estimated at four hours in summer; but if the plovers are frightened off for that long, their eggs may scald. Conversely, people who come to the island at night to fish may also scare the birds off the nests, exposing the eggs to temperatures too low to allow chicks to hatch.

On Monomoy, the peril of human disturbance has an added, harsh twist because the gulls are always standing by to exploit any vulnerability. Once the plover nests are undefended, predatory gulls move in—to feed on the eggs and, later, the chicks. Short of intervening to reduce the gull population, wildlife biologists and managers have had little room to maneuver to give piping plovers a better chance at survival and reproduction. And, these scientists are quick to acknowledge, gull control is difficult, at best, and represents a human intrusion through management that carries its own set of perplexing questions and problems.

TINKERING OR NOT, blaming or not, no one is suggesting that it is an easy task to be a spectator at the death vigil of another creature.

I learned that lesson as my last stay at Monomoy Light was coming to a close in the summer of 1995. As a consequence of nothing but chance, I spent the better part of an afternoon attending to a dying gull. The next day I buried it.

I had set out from the keeper's cottage, heading northeast to reach the ocean. I was emerging from the dunes along the sprawling concave of a washover fan emptying onto the beach when I saw it—a herring gull, flapping awkwardly, flailing in the swash. It lifted its wings, thrashing but going nowhere, as the waves swept up and threatened to carry it off into the surf. But there was still life enough left in it for the bird to struggle against the sea and await its death on the sand.

I could tell it was nearly finished. It lay on its belly, its head down, the beak resting on the sand, the bottom bill tipped with a spot of red like a drop of dried blood. I was able to approach it, stand over it; and though at first it tried futilely to lift its body from the sand, it soon surrendered and lay still, its eyes half closed and vacant.

It occurred to me that there would be a kindness in killing the bird, rather than leaving it to die slowly in the surf, but I couldn't bring myself to the act. Instead, I sat six feet from the gull, talking softly to it for a while, then humming, all the while following the learned impulses that have become part of my own being—watching intently, taking notes on the bird's last leaving.

Every few minutes for nearly half an hour, it would open its beak, as if to issue a trumpeting call, but no sound came—only the action of choking. When the violence of that would pass, the bird would make another hapless attempt to fly, raising itself to its large, webbed feet, and flapping wildly, each try ending with its head dropping forward limply, its beak probing deeper and deeper into the sand.

If I had not been there, perhaps nothing would have bothered the bird in its final strife. Overhead, the sky was empty of sun and other gulls, and dark clouds tumbled in out of the north, the color of the approaching storm corresponding perfectly with the sooty feathers of the gull's mantle and wings and the dark, insignificant moment I was witnessing. A herring gull was dying, that was all—a big, common bird, its white head trembling with the palsy of its passing, while in the waters offshore, harbor seals bobbed and dived—six here, floating by; then two more; and, at last, one alone, rolling without alarm in the waves. Even I left for a time, sensing the death was hours off, at least, and wondering, after all, if my presence might not be more disturbing than comforting to the gull. I promised to come back later in the afternoon and bury it, not knowing why, but telling myself it was a gesture worth seeing through to the end, since death had made us intimates for the afternoon.

I walked off into the path of the storm, toward the far end of the island, going just to be able to say I had reached the destination I had set for myself, and then returned. The heat of day had dropped away, the beach getting colder and colder; and finally, the sky let go, and it began to rain. By the time I made it back to the gull, the storm was at its full, the downpour hard, great sheets of rain being pulled, one after another, like shrouds over the sand. But the bird was still alive, barely. And whatever it takes to move someone to hurry death in the name of mercy, I had not discovered it on my long walk. I stood in the turbulence and told the bird I was sorry—but I did not know what for.

I left it to die alone.

The next day I returned and found the carcass, as I knew I would, worked over by some other creature—another gull, most likely—the heart ripped out of the breast, a gaping hole, but otherwise intact and, at that, an impressive form, even emptied of life.

I dug a shallow grave for the remains, lifted the gull one last time into the air, this time unnaturally, with its pink feet to the clouds, its slack neck and exhausted wings unmoving, and laid it down in the sand. Following the Native American tradition, I sprinkled tobacco over its body, covered it with sand but no words, placed ten rocks in a straight line down the center of the trivial mound, and placed at its head a piece of driftwood that curved at the tip like a talon.

No one is going to miss a gull, I thought as I climbed the dune to the light, some slight fluttering in my chest as heavy as stone, but no more sentimentality stirring there. Still, there was a sadness I could not shrug off—not the death, but the isolation of it. That, and the heart—the harsh way it was taken.

I remember how that feels.

WHEN ALL THE HUMAN DEBATE HAS RECEDED, it still seems that the life of a bird, compared to human existence, is clearer, somehow, and more predictable—its imperatives determined by the familiar cycles of migrating, establishing territory, courting and nesting, shedding and renewing plumage, and being funneled back finally to the flyway that will return it, and its offspring, to wintering grounds and in spring to the place where they will start again. Thus they move through the year and the sky according to a map already in their blood.

It was daunting to probe the minds and behaviors of the birds during those late summer and early autumn days on Monomoy, when the air was so full of stirrings. Even years later, back on the mainland, I could not see a flock of birds making a condominium of the canopy without being reminded that there is *being* and *knowing* that sustains life and is beyond me in all the ways that matter.

I try to keep that in mind, emulate the migrants' sense of purpose as best I can, and keep my eyes and heart as open as the outstretched sea. I never know, really, when or what nature has in store to set my thinking straight. One night the moon showered down over me through the darkened, inland trees in a patchwork of light. Until I noticed it, I had been thinking, oddly, all morning, that my business was important, imperative even, to finish before day's end. Then I saw the blackbirds gathering and Canada geese arriving—all of them filled with departures and high purpose. From my little, low world, I knew the birds traversed highways not my own and carried on without concern for human hurry—we predators in steel boxes, hurtling down concrete pathways, governed by landlocked lights that blink from green to yellow to red. The birds move, and cease, according to other, more comprehensive commands, and in their flight I have found my human character both dwarfed and dignified. I remember again how small I am, and in right relation, how precious—like them—a part of the exquisite, intricate balance of life.

Everything, and everyone, has an appointed course to keep. I cover the recognizable ground of my own existence with more care for having witnessed the birds piloting themselves to Monomoy and on farther, down the planet. In my own dumbstruck awe, attention and a slowed gait are the best my limbs, working in concert with my mind, can do on a shining night in early autumn.

But in my mind, from memory, I would give you the hawk.

The northern harrier—*Circus cyaneus*—haunts the shallow ponds and marshes stretching from the lighthouse to the Sound. During any given day, it floats on the air currents as gracefully and effortlessly as the scoter and oldsquaw ride the waves offshore.

It became the bird of Monomoy to me, even though there are many rarer and more spectacular birds on the island during fall migration. The white-rumped harriers skimming the perimeter of my vision during those early island days were mostly young ones, those described by the Peterson guide *Eastern Birds* as being "rich russet below"—the chestnut red exposed like the suggestion of a setting sun over the waters as the bird makes its low-altitude reconnaissance for prey.

One day during my first stay, shortly after noon, a male soared by as I sat on the deck trying to write, my chair situated so I could keep an eye on the marsh. It flew within fifteen feet, rolled over in the air, its wing tipped in black.

The marvel of the hawk for me, in my human envy, was its easeful purpose, its untroubled, deliberate flight. I had watched, with admiration, as the kingfisher hovered for ten seconds or more, as if suspended in midair, over the pond; and I had tallied the dowitcher and the whimbrel watching over the Powder Hole. But my heart rode on a harrier's wing.

The summer was coming apart at the seams on Monomoy, and fall was everywhere. Some of the signs were absences: The egrets had abandoned Lighthouse Marsh, and although I saw fourteen congregated at the Powder Hole late the previous week, ponds there seemed vacant and dull without them. All over the island, the teal and black ducks, mallards and Canada geese were jamming the waters. The songbirds had flown, although a few stragglers—sparrows, mostly—were hanging on.

Mornings, I sat on the deck, wearing a ski vest or a winter jacket, huddling alone in the wind, dreaming of the long flight, the birds in my mind like augurs—but of what?—the cold dawns to come. I had met the meadow vole, his underbrush passages leading at last to my back door. One morning we sat for several minutes, regarding each other, keeping our distances, while he nibbled bits of rice cake that had fallen to the sand.

They come and they go, the creatures pacing out their days. I had walked within a stone's throw of deer in every part of the island and watched them disappear, lying down in the tall beach grass, swallowed up in the camouflage of the great green and yellowing waves.

If I could, I would offer them up, a blessing for you; the toads tricking the eye into seeing the ground swell; the terns, silver in flight over the platinum sun-split waters; the sky a dome of delft blue; the stars a needlepoint of light in the late hours. And as another, later autumn dissolves now into silence, canopy upon canopy of peace drapes over me: the trees rattling their dry bones; the clouds of geese silhouetted against the moon, pulling the flowing veil of travel over the night sky; the stars spilling like lost hopes—insignificant and shining—through the heavens. I take one more deep breath and let it all fall away, the seasons gone, the secret dread, the messages in the brown and golden landscape, all the known, obscure refrains of natural song.

For I am only one more creature coming and, soon, going away again, making my own migration. Here on a delicate, durable spit of sand jutting into the ocean, protecting a continent, as a rib buffers the breath, I have learned my small place and know it large enough for life. In my time, I have become part fur, part feather, part stone.

I, too, contain the memory of the tern's tremendous, shimmering flight, and, fired by imagination, I coax from it an internal flame. In the shadow of that light, I decant the stories older than time—how those first wings and fins over eons changed to fit their task; how the ice helped, and the winds and waves; how each voice, each voyage, had its own cadence and destination. Because of the courage and willingness of the birds to light on an impossible journey as though it were nothing remarkable, I will pipe for them, with them—till the last breath carries us on elsewhere in the hymn and pulse of irrepressible life.

Assembly of Seals

In focusing on the nature of the varying relationships that can develop between humans and animals, one is led inevitably to the question: what actually is an animal or a human?

—P. J. Ucko, *Signifying Animals*

THE THIRD DAY OF MY RETURN to Monomoy, I spent the afternoon singing to a seal.

It was not the first unforeseen event of the day. Water—the element on which both humans and seals are dependent for life, and seals for habitat, as well—had been the subject of earlier unexpected news that morning. I had been waiting for results of a test of water I'd taken from the old pump in the keeper's cottage a week or so before my scheduled stay. I'd turned the sample over to the county health department and portaged in enough water for the first few days of the trip. I had assumed the pump water was not fit for drinking. Under ideal circumstances, I'd figured, I could use the water to wash dishes and fill the solar shower for my own bathing. In fact, I'd just finished showering in pump water warmed by the sun, remembered idly that I hadn't heard yet about the water test, and decided to use the cellular phone to call and check.

The health department official I reached told me in no uncertain terms that the water was not fit for any use—drinking, cooking, or washing. That assessment cast a pall over the idea of staying on the island for more than the few days for which I had enough water. We spent a few minutes mulling over the possibility of chlorinating the pump and pipe—enough, at least, to make it possible to bathe in the water. But without a second sample, we couldn't be sure that technique had worked. We finally settled on a ratio of chlorine bleach to solar-shower water that would be sure to kill any bacteria, and that seemed more hopeful news, at least for personal hygiene during my stay.

We said our good-byes even as I was reluctantly hatching a plan to hike out to Nantucket Sound—significantly warmer and less breezy than the ocean side of the island—for a bracing saltwater bath to remove any harmful bacteria or sediments I might be wearing from my earlier shower. The idea of lounging in the salts of the Sound would have seemed glorious in mid-July, but surrender was the best I could muster in early June. On Monomoy, as all over the Cape, June weather is consistently cool—and in the keeper's cottage, still dank from being boarded up all winter, it was cold. Still, tainted water can pose grave dangers to living safely indoors or out, and I knew I had to take the issue seriously. I grabbed a bar of soap, shampoo, and a towel, resigned myself to the chill of the air and water, and set off down the trail leading to the open beach for the fastest toilette I have ever accomplished. By the time I was striding back, the cold wind was drying my hair and I was using the towel for a scarf.

When I reached the keeper's cottage, I was still too chilled to sit in the fusty rooms. I made a quick pot of coffee, and drank two mugsful just for heat, cradling the cup in mittened hands. As soon as I had downed the brew, I moved on, deciding a walk out to the ocean would provide exercise enough to warm me up.

I half jogged over the banks of dunes, up and down through the sand, ducking the dive-bombing of gulls to reach the outer beach. I passed the grassy gulls' nests with buff and darker brown spotted eggs, sweeping up ticks on my jeans as I walked, until at last, I cleared the last crest to the beach. About twenty yards off to the north, something dark and still lay on the beach. I wasn't certain at first what it was—debris spewed from the sea can look like almost anything from a distance on an empty ocean beach. Driftwood takes on the appearance of bones; a mat of seaweed can seem a twisted towel or blanket from far off. I didn't know what had washed up on the berm, but I was still feeling out of sorts about the contaminated water and was not inclined at that moment to put a happy spin on things. I guessed the motionless form to be a seal, but I wasn't sure I wanted to see it if it was dead and decomposing.

As I drew closer, I could tell the shape at the water's edge was a seal—almost certainly an immature harbor seal, judging from its size and features—lying on its side. Because the wind was strong and out of the northeast, the sound of my movements over pebbles and sand spread back into the dunes, and the silvery, spotted seal never heard me coming. I walked slowly, with care and hesitation, toward it, all the while expecting either to discover it dead or to waken it and send it fleeing into the surf. Finally I was standing less than ten feet away from the mammal, my beige canvas jacket slapping like a sail in the wind.

The seal, still unaware of my approach, came slowly to life. It gently rolled over on its back, rubbed its eyes and smoothed its whiskers, and let its pawlike flippers fall to its sides. Then suddenly, as if oddly sensing my attention more than being alerted by hearing or smell, it rolled its head back, blinked, and stared my way. But beyond that, it did not stir.

For a few minutes we eyed each other, moved but unmoving, the seal waking, I bristling with marvel at the animal's proximity and its unruffled presence. For a few seconds, my body flared with a rush of curiosity and fear—just what the seal, I imagine, must have risen to in waking to my eyes and the afternoon sun. But fascination was the greater part of what held me, and for an instant that seemed to jettison us out of time, the seal and I stayed, transfixed and befuddled, our gazes locked, one to the other, across the abyss of differences between species.

Then my primate consciousness, stuck on survival and making sense of things, receded once more into itself, and I began to think human thoughts, with words and intent to them. My whole idea was to cling to the moment, to the unexpected connection, and it occurred to me that my stature must seem an unfair advantage to the seal. Slowly, slowly, I lowered myself to the sand, until at last I, too, lay on my side, propped up on one elbow. There was no unnatural sound, no mammalian howl from either of us, only the wind, the breakers, and the agitated gulls overhead.

The seal and I remained like that—suspended—for the brief duration that can seem a long time when prey and predator are rendered instantly insensible by unexpected contact—not seconds in this case, but minutes. The quiet and harmony held on, pierced only twice by a low, murmuring *gaooooh* and *ahhooooh* from the seal. Once it hissed and, later, let go the sound of a guttural bark like a cough or a throat being cleared.

Then the ocean introduced a change in our positions. A wave broke and washed high on the sand, sloshing the seal, which lurched out of reach of the water but nearer to me. When it came to rest again, its head was only a foot from my belly, my body coiled like a crescent moon in the sand above its snout.

I had never been so close to a seal, and the experience of having an ordinarily distant mammalian relative suddenly so near made something give way inside me, like a twig snapping or the catch of breath that attends delight. The seal, with its round eyes and impenetrable stare, did not attempt to escape, and I did not try to touch it or to recoil. Instead I lay there, took in every inch of its amiable face, its quill-like whiskers, its silken, dappled fur, trying to record the details and savor the moment. Whatever fear I first had felt ebbed away.

Somewhere in the back of my mind, I was calculating the encounter as a zoologist might, running an internal reel to capture all the identifying characteristics of the animal—its shape, size, and color; the condition of its teeth and coat; injuries sustained and scars healed in the wild; its behavior and, to my mind, its mood. But the more dominant part of me just then—that unbroken creature within who still opens unwary to the world—had other designs. In the middle of my reverie with the seal, not knowing just what to do, I began to sing softly, as a child might to a bird, as though peaceable sound alone could hold us together. I wanted to offer the seal *something,* some reassurance, a gesture of trust, to announce that I was not there to hunt or to hurt, but only to witness and observe.

With history as evidence, the seal should have been wary of any human. For the better part of a century, from 1888 to 1962, Massachusetts offered a bounty on harbor seals, ostensibly to preserve fish stocks from depletion by any species other than humans. By the early 1900s, these seals were almost exterminated along these shores—and there was little change in the number of fish. The bounty on the seals was finally lifted in 1962, and federal protection came ten years later. Though humans came close to eliminating harbor seals along the Massachusetts coast, human restraint—in the form of simply leaving the seals alone to fend for themselves—has brought them back. Since the midseventies, populations of harbor seals have nearly doubled off southern Massachusetts.

The seal lying with me did not seem alarmed, and I was intent on keeping the calm between us. But even as I try now to give account of the moment, I am imposing on the interaction too much deliberation. For at that instant, reclining in the sand, with a seal almost in my embrace, I wanted only connection without touch, alliance without fear. Not knowing what the seal understood as sign or language for "safe," I simply followed the same logic I had as a child, when singing made me feel less afraid. The first tune that rose in my throat was "You'll Never Walk Alone," a well-worn inspirational song I learned in grade school and often hummed when walking the alleys of my Chicago neighborhood on the way home from classes. Now, with the seal, I murmured the familiar "walk on, walk on" refrain and tried not to screech on the soprano notes. I had never managed to croak out much above a high tenor or low alto, though, and when I struck the more challenging notes, the seal cocked its head from side to side like a dog. At that, I figured it would be best to try a different tune.

Still unthinking, I heeded what seemed to come by instinct, and from some atavistic choir in me rose the words of the hymn "The Church's One Foundation," erupting from some cloying Presbyterian past. I managed to deliver two and a half verses before the seal and I both seemed spent by the burst of piety. I resorted at last to something more contemporary, a Wynonna Judd country song, "My Strongest Weakness." And with that, the seal took off, belly rolling into the surf, where it floated, just out of reach, bobbing in the swells, spying me.

I stood up, brushed the sand from my clothes, and walked off down the beach for a mile or more, the seal staying with me all the while. Even when I watched it dive out of sight and remain submerged for a half minute or more, it was never really gone. I tried once more to beckon it to shore with a Scottish folk tune sung to seals off the coast of the Western Highlands, but my rendition produced no effect. The seal off Monomoy clung to the sea, diving and then reappearing, just parallel to the point I had reached in my walking. When it had taken enough time for a good look, it would duck under the swells again, amusing itself with me, from a safe distance, for nearly an hour. Then, as suddenly as I had arrived on the beach to interrupt its nap, the seal disappeared for good, leaving only the thin sound of a human song drifting harmlessly in the wind.

To encounter a seal at extremely close range is to understand the appeal of human constructs like "collective memory" and "the unconscious." Part of the prompting may derive from the gaze of deep and intimate innocence that a seal communicates, or its sleek fur and gently curved trunk that appear oddly like the human form—especially among North Atlantic seals, which do not grow to the strapping hulk of walruses or elephant seals. In any case, we have long thought of them as being like us.

The marine mammals—including pinnipeds (eared seals, walruses, sea lions, and true seals), cetaceans (whales, dolphins), sirenians (manatees and dugongs), and sea otters—are believed to have descended from land mammals that had abandoned solid terrain for the sea many millions of years ago, probably because the early oceans offered escape from certain predators and the prospect of abundant food. The cetaceans have lived longest in the oceans, evolving and diverging from a primitive hoofed mammal between twenty and sixty million years ago. They and the sirenians evolved from the ancestor of present-day elephants and aardvarks, and accommodated themselves completely to the new environment of water, giving up hind limbs for a tail to aid in swimming and diving. Sea otters are considered the "youngest" or most recently evolved of the marine mammals, having taken to the water between five and seven million years ago. Over time and through adaptation to life in the water, they have retained their hind feet as modified flippers and kept forepaws approaching those of carnivores on land. Among the marine mammals, sea otters are the slowest swimmers and the least nautically sleek, and they may come onto land for very brief periods. But their lives are lived almost exclusively at sea; even mating, breeding, and raising pups take place in the water.

Seals, as pinnipeds, are "fin-footed," "feather-footed," or "wing-footed" mammals, depending on how the Latin root of the word is translated. They are linked in evolution, some thirty million years back, to a meat-eating land animal, but scientists have not settled the question of whether there is a single, common line or two distinct paths of evolution. In either case, however, one ancestor seems to have been similar to a bear or dog. Even today, the skulls of sea lions, bears, and wolves can be difficult to tell apart.

According to one theory, these large, primitive forbears of seals shunned land for the seas as the climate changed and northern waters, along what is now North America and Europe, became cooler and rich in nutrients. They evolved into the exceptional swimmers and divers we know today as seals—fur seals and sea lions in one family, *Otariidae* (distinguished by their external ear lobes), and true seals, *Phocidae* (with ear holes rather than lobes), in another. Walruses form a third distinct family, *Odobenidae.*

For every adaptation that ameliorated a seal's life in the sea there was a corresponding change—and often, loss—in their capacity to survive on land. Clearly, the transition from paws to fins, or "feather feet," eased the seals' movements in water, better equipping them for speed and rudderlike control of direction. But fins were not particularly good for use on land and made for cumbersome escape from predators on rocks and beaches. Likewise, the thick layer of blubber they developed enhanced their survival in the cold waters of the oceans but was a potentially fatal burden on land. For a marine mammal, overheating can mean death.

Most seals, despite the nickname "sea-dogs," do not engage in the canine behavior of panting to cool down. To survive the heat, seals—like marine mammals generally—possess a network of intertwined veins and arteries in their flippers, flukes, and dorsal fins, where there is significantly less blubber than in the rest of their bodies. This system works both to cool and to warm their bodies, under whatever conditions they find themselves, by causing excessive heat or cold to dissipate from their blood as it is pumped between the heart and extremities. In seals, this network is especially intricate and exists even in the superficial layers of skin, which allows blood to surge rapidly just beneath the surface, speeding the loss of heat or the relief from cold, as needed. The use of this network of blood vessels and its capacity to regulate heat is what is at work when seals, lounging on the shoreline on a hot day, burrow into wetter, cooler patches of the beach, wave their flippers in the air, dip them into cool, wet sand, or flip sand onto their backs for protection from direct sunlight or scorching heat. By the same token, on a bitterly cold day, hauled-out sea lions in the southernmost reaches of the sea will tuck their flippers under themselves to conserve heat. Some species may huddle together onshore to protect themselves from cold, particularly if there are vulnerable young. In frigid water, fur seals and sea lions arch their bodies into a jug-handle position so that both hind and fore flippers are above the surface, staying drier and warmer than the bulk of the body.

Adequate food—and more precisely, sufficient fat—is essential to seals because the ocean in winter and summer can be a rugged environment in which to stay warm. All seals but the largest—walruses and elephant seals—are covered with hair, or fur, which insulates them to a degree on land, but it is inadequate protection when the animals are wet or are in water. For those conditions, a hefty layer of blubber is more reliable.

Many seals hunt at night and sleep on the beach by day, partly because of their internal thermostats. Seals in seawater lose heat at a rate twenty-five times greater than when they are on land. In winter, when you might expect the thermal tables to turn, seals still are likely to haul out on land just to bask in the heat of a diminished sun. Temperatures must dip below freezing and the wind must pick up to more than twenty miles an hour before it becomes more prudent for seals to remain in the water for warmth.

To bridge the two worlds of sea and land, seals have evolved complex adaptations for breathing. They are impressive divers—the gigantic elephant seals are capable of plunging to nearly 5,000 feet below the surface and staying submerged for more than an hour. North Atlantic seal species generally stay at much shallower depths and surface within a few minutes at most. But none breathes underwater; all exhale before diving, forcing oxygen out of their partially flattened lungs to ease both their descent and their rapid resurfacing. Whereas a human would find the air pressure in the sinus cavity painful during an unaided dive or would experience "bends" in ascending too quickly to the surface, seals have special adaptations for diving and resurfacing without discomfort. First, they have hollow spaces for air only in the less sensitive lungs and ears. Additionally, they have nearly collapsible internal organs and airways to reduce the buildup of internal pressure. Even the trachea canbe compressed, according to the marine biologist Marianne Riedman.

A seal's circulatory system and the blood itself are uniquely suited to the task of retaining oxygen under water. At the same time, seals' systems conserve oxygen by feeding blood only to essential organs, slowing the heartbeat and metabolism and reducing body temperature. Even the muscles, deprived of oxygen in diving, are specially adapted for the hard work of submersion: They contain a special iron-bonding pigment that stores oxygen. And seals, like whales, have an automatic internal mechanism that cuts off their breathing if they should be knocked unconscious during a dive, thereby preventing drowning.

Among the remarkable everyday feats of seals is their reflexive act of sleeping, which can occur on land, on the surface of the water, or beneath it. Research suggests that, like humans, seals dream, but solely on land. Rapid-eye-movement (REM) sleep, during which dreaming happens, can take place only as long as the mammal is not underwater. On occasions when harbor seals—frequent visitors to the coastal waters all along the northeastern United States—snooze in the water, they tend to "bottle" in an almost vertical posture, with only the face, snout, and nostrils breaking the surface like a snorkel. This position enables regular breathing during sleep.

One of the greatest challenges of sea life for mammals is adapting to the saltwater itself—drawing water to prevent dehydration and getting rid of the salt. Pinniped kidneys are uniquely fit for flushing salt from water that has been drunk or absorbed by the animal, and so much salt is excreted through the urine that there is a net gain in water to the animal's body—just the opposite of what occurs in the bodies of land-locked humans. Still, most of the fluids that seals need come through fish they take as prey.

A seal's life is occupied primarily with the essentials of survival for the individual and the species—namely, everyday feeding and safety, as well as the overarching matters of mating, breeding, birthing, and raising young. Survival for animals bound to both land and sea requires a full range of mammalian senses and skills, and then some. Water, air, and land all pose particular threats or provide unique protection, and seals have evolved to capitalize on the benefits of each medium. For seals, as for all creatures in nature, necessity is the mother of evolutionary invention.

Scientists have learned, for example, that seals hear quite well—but better underwater, where they get most of their food, than on the surface or on land. Compared to humans, seals are able to discern a wider range of sound and frequencies under water, though on land that capability is dulled to a degree, and it is believed that they hear less well on land than most carnivores. Partly this is because they have evolved away from the large external ears of land mammals. The ear holes, or small external ears of some species, aid them in diving and underwater existence generally, but they detract from land-based hearing. On the other hand, although seals appear to have little to no ability to smell or taste underwater, on land they seem to have an acute sense of smell, which is essential for certain rituals of reproduction and for protection. Under ordinary conditions, they can smell the presence of a human hundreds of yards away.

Submerged, pinnipeds outstrip humans in hearing, but they are outdone by the cetaceans. Seals lack the apparatus to deliver the keen hearing and sophisticated vocalizations of the echolocating whales and dolphins, for example, which emit a complex series of clicks from the skull to pinpoint, investigate, and perhaps even stun their prey. In fact, no one knows for sure whether—or how—seals might also use a type of echolocation; marine biologists still are studying this question. Gray seals have demonstrated their inability to use sound to locate themselves and navigate. And harbor and ringed seals, along with some sea lions, have been observed sounding with underwater clicking, particularly in murky waters, which suggests that that they are doing something like echolocating, but just what that something is remains unclear.

A seal's vision, too, must be extremely sensitive under a wide range of conditions. Seals must be able to see well under and out of water, in habitats shaped by bright snow, shining sands, and the low light of rocky coasts in storms. Consequently, everything about their eyes—from the size of the cornea to the structure of the outer layer—is designed to enhance and protect vision in water and air. From where we sit, those eyes seem especially large and dolorous, but the size is directly related to the versatility of their vision, the bigger to see their prey—and us. Their eyes are equipped with lacrimal glands for tearing which cleanses the surface of salt and sand, and they have inner eyelids that protect the eyes from blowing sand and drifting debris.

The tactile senses in seals reside in the skin, and the need for touch for warmth, recognition, and socialization in some species may explain the behavior of individuals huddling tightly together or, in some cases, piling atop one another on rocks and sandy beaches. In addition, seals possess what we think of in land mammals as whiskers, and these are highly sensitive. These whiskers—or vibrissae—contain up to ten times as many nerve fibers as the whiskers of a land mammal. They are used for navigating and foraging, and in much the same way that the position of a dog's tail communicates fear, play, or aggression, the position of a seal's vibrissae is a clue to its mood, its state of attention or aggression.

Humans tend to see seals poetically and imaginatively as among the gentlest overseers of the temperate shores, but they—like their marine mammal relatives, the whales and dolphins—have developed a very complex social dynamic that can involve fierce contests for power and the right to reproduce. Eared seals, including sea lions, form harems, with a dominant male, or bull, who takes charge for about two weeks—before being challenged and potentially vanquished by a competitor. The spoils, beyond the harem's females, include the territory of rocks on which to sun.

In frigid environments, such as Antarctica, these contests for control can be lethal; an open wound generally spells death because of the extreme cold and unrelenting winds. There, because of the high stakes, southern elephant seals that wrangle for access to females often, though not invariably, stir up conflicts that are more bluster than outright battle. And as the bulls age, squabbles become mostly posturing, slashing, bumping, and noise. Even so, over time the chest of a combative male begins to resemble a shield of scar tissue—witness to his stubborn persistence in the annual battle over mates.

All young seals are born on land (or ice), but mating and breeding behaviors vary from species to species and are linked inextricably to where and how many females are available. This makes the process of reproduction in part a function of place: Pinnipeds tend to breed either on land, on pack ice (floating ice), or on fast ice (attached to land).

Some species copulate on beaches or rocks while others—particularly those who use ice for haul-outs—may be forced to breed in the water. Many gather in colonies, males overseeing and mating with whole harems of females; but in some instances, where the opportunities to mate and pup are short and the environment is unstable, monogamy or serial monogamy is the rule. Among species that mate on pack ice, for example, males do not have ready access to a number of different females.

Harbor and gray seals are the most common species on Cape Cod and nearby islands, and harbor seals return north year after year from May to late June to the same island sites in Maine and eastern Canada to mate and bear young. The female bears a single pup, which is capable of crawling and swimming at birth. Generally, however, the pup will remain with its mother to nurse for at least four to six weeks, and perhaps longer, to ensure that it will be ready to survive in the sea on its own. Its first food is its mother's high-calorie milk, but it quickly heads to the water to feed on schooling fish, and eventually it will cultivate an adult diet of squid, invertebrates, and larger fish.

Biological and social evolution has enabled seals to survive far and wide over the oceans of the planet. Their range extends from the polar extremes of the Earth to the regions with more moderate climates. The ferocity of their behavior, such as posturing and fighting among males before breeding begins, has evolved to suit the prevalent conditions in which they live.

Whatever their ancient past or their adaptations to the double life of a land- and sea-dwelling mammal, seals throughout human history have struck a responsive chord in human consciousness. We, mostly shorebound, not only feel a kinship with our water-world cousins, we also share with them such characteristics as breathing with lungs and bearing and nursing the young. And, for all our obvious differences, we and they seem unwilling and unable to shake off the charm of that innate, enchanting curiosity that seems a shared legacy among many seals and humans, especially children.

"There's a connection between people and seals," says Bob Prescott, a wildlife biologist who has studied seals on Cape Cod since the mid-1970s. "There's an expressiveness we see in their faces . . . the winsome look of the round eyes . . . the big brown eyes. They do react to our presence. They are keenly aware of us being there." The director of the Massachusetts Audubon Society's Wellfleet Bay Wildlife Sanctuary, on the Outer Cape, Prescott lectures on marine mammals as part of Audubon's educational outreach programs around New England. His experience is that most seals, if approached by humans, will not react as the seal did who stayed near me for nearly half an hour on the South Monomoy beach that early June day. They are far more inclined to take refuge in the water, retreating twenty-five to fifty feet offshore. At that distance they feel safe enough to give play to their curiosity, and often they will drift along, as my companion did, keeping pace with humans walking the beach. Or they will dive and resurface close to beach-combers or fishermen. Seals have been known to swim within ten feet of shellfishermen dragging quahogs from the sandy bottom of Nantucket Sound along South Monomoy. Then, for an hour or more, the seals bottle with only their heads above water, watching the humans at work.

Seals—especially those visible along the shores of the North Atlantic—are as inquisitive about us, it seems, as we are about them. Of all the marine mammals, they appear, at first glance, to be the most compatible with humans. Their large round eyes give them an expression of innocent empathy, and their behavior for the most part backs up that impression. They tend not to be aggressive unless threatened or cornered, and at times are more tolerant of human presence than many animals. Certainly, seals approach us more often than, say, the cetaceans, no doubt principally because they are better suited to join us on common ground—in air, on land. Seals can survive—and, in fact, require—long periods on land as well as in water. Given these common needs, it is hardly surprising that humans, over time, have imagined some vague, shared mammalian memory between our kind and theirs, and have celebrated, in verse and song, the eerie sense that we can meet on the edge of the sea and plumb the watery unconscious to bond with them.

I felt the hypnotic urgency of that fantasy—and longing—that early summer afternoon on Monomoy: It was impossible to maintain fear under the guileless authority of the seal's gaze. Rather, it disarmed me utterly and aroused in me a feeling of enchantment, of being linked, magically, with an Other from the oceans. Back home, inland, under incandescent light rather than the ivory shimmer of the sea, such emotions might sound preposterous. But humans have always sensed a presence more ineffable than animal in the seal, and profound attachment to these creatures has a long, venerable history among human populations living along the ocean. "I am a man upon the land, I am a selchie on the sea," goes a Celtic folk song. Many of the ancient tales of seal-and-human bonding still persist today in the British Isles and Scandinavia, where seals are commonly found inhabiting the same shorelines that humans haunt to be near the best coastal fishing. Over time, in fact, seals—which are superior marine predators—became linked in legend to the fate of fishermen and their catch. Seals, and even their pelts or other parts of the body, had to be treated with respect and were believed to bring good fortune in fishing.

In Icelandic myth and lore, seals—like humans—are among the few beings that are able to mediate between land and water. Seals, however, in some sense possess a quality that people need to survive and, therefore, must court from the sea mammals: power over fishing. The animal's unique abilities to flourish in water and on land led to a belief that, hidden beneath their outer coats, seals had a fully human form and nature. If a successful hunter managed to kill a seal and offered one of its eyes to a less proficient hunter, that man's luck was certain to improve. But other fables warn of ill-fated crews who are drowned at sea after attempting to shoot a seal. Whatever bond existed from generation to generation between humans and seals, it was one that remained beneficial only to those who regarded the seal with reverence for its real and symbolic power.

One of the most primitive legends of the seal, according to Sir James George Frazer in *The Golden Bough,* claimed that the fur of the marine mammal stayed miraculously animate even after the animal was killed and stripped. The sealskin, it was said, "remained in secret sympathy with the sea" and on its own would ruffle during the ebbing of the tide.

Many fables related to the seal are tied to the animal's skin—both its extrinsic value to humans who used the pelts for clothing and boots and the blubber for fuel, as well as the intrinsic, metaphysical worth the skin was believed to have for traversing not only between sea and land but also from the material world to the spiritual. Such stories are more easily understood when regarded from the viewpoint of people whose lives and livelihoods were enacted on the sea and at its edge, where encounters with seals were commonplace and hunting the animal was a well-integrated part of communal life. Worldwide, similar magical qualities are attributed to animals stalked by indigenous peoples for food or clothing. And as is frequently the case in legends across cultures, the bond between human predator and animal prey is intense and intricate, expressed in complicated rituals and beliefs, not the least of which deal with the treatment of pelts and the use of meat.

The native peoples of the Bering Strait regard the bladders of marine mammals, including seals, to be the residing place of the animals' souls, and should therefore be handled with respect and preserved with a sense of awe. "At a solemn festival held once a year in winter," Frazer recounts in *The Golden Bough,* "these bladders, containing all the souls of sea-beasts that have been killed throughout the year, are honored with dances and offerings of food in the public assembly-room, after which they are taken out on the ice and thrust through holes in the water." Such ritual acts of returning specific animal parts to the waters in which the creatures live not only serve as a kind of communal recompense for the hunt, but also express a deep-seated belief that such acts have the power to restore the hunted "herds," and thereby assure the fate and safety of the human community. To act in a way that dishonors the animal is to court disaster on the sea during the hunt. And it could have calamitous consequences on land, where shape-shifting seals might be abroad, mingling with humans and dispensing a sort of natural justice designed to reestablish a sense of balance between the human and nonhuman worlds.

The miraculous nature of the seal and its enchantment over humans transcends the hunt, however, gathering the creatures of sea and land into a vast net of metaphysical meanings and correspondences. The survival of Inuit (Eskimos) in Arctic coastal settlements was inextricably tied to seals in every facet of individual and communal life, as depicted widely in their preserved art. There, as well as in China, there is evidence of the conviction that seals could somehow transmit not only vitality but also potency to humans. The Inuits gleaned from the walrus its three-foot-long penis bone to use as an *osik*, or walking stick, and in China the genitalia of breeding seal bulls were harvested, dried, ground to powder, and used as an aphrodisiac. Certain clans of the Hebrides trace their ancestry to seals, while in Irish folklore seals take on lives and capabilities remarkably like their human counterparts. Most notably, seals can change form and take human lovers, or otherwise perform miraculous acts benefiting human beings. Certain Celtic tales relate accounts of seals rescuing and nurturing abandoned infants until the children could be reclaimed by or returned to their human families. Other stories tell of the "sea trow," "selkie" or "selchie," or "silkie," a seal, typically female and usually gray, that is capable of "shape-shifting," or transforming its outer appearance and living among humans. In legend, these creatures, bridging the worlds of water, land, and air, would come ashore at certain times, such as on full-moon nights, slip from their glistening skins, and dance on the rocky beaches. A man who managed to gain possession of the skin of one of these seals held the silkie captive, often even as a wife, until the seal maiden could recapture her skin. Such seal women, it was said, could be recognized by their rough hands, a slight webbing between their fingers, slow breathing, a love of diving and swimming, and their skills in the arts of medicine, midwifery, and forecasting the future. And silkies might also be found out from a strange, lingering quality of their sea life: When these changeling beings arose from sunning on the beach, they would leave a damp imprint glistening with salt crystals, even if they themselves appeared dry.

These stories underscore the sense of a deeper kinship than the merely human to human, a spiritual bond that unites all of life and both defines and transcends the limits of time, space, and physical form. On even a mundane level, these legends reinforce a system of values that elevates human obligation to the level of participating in, protecting, and preserving the environment. At their most basic, these stories teach interdependence—that it is not in our interest to act with disregard, or even dispassion, toward the creatures that occupy the planet with us. Further, they remind us that other animals, and spirits, have interior lives and faculties that we ignore at our own peril. We are not alone, they say, and ought not think, fear, or behave as though we are.

On a strictly practical level, since prehistory the link between seals and humans has been forged by much more than fancy. For thousands of years, humans hunted seals in the Arctic, and their motivation was survival. They needed the meat for food; blubber for oil and heat; hides for boots, clothing, and the coverings for skin boats; rawhide for laces and harnesses. And much later, seals' coats provided the foundation of a fur industry in Europe and America that decimated many arctic herds. Seal hunting persists in parts of the world today, and the protection of some species remains uncertain.

Regardless of the ebb and flow of public opinion and the vagaries of human law related to seals, the bond endures between us and them. It probably is not to their benefit that we find them to be so like us—linked even more subtly and durably to us in ancient myth than even our everyday cohorts—dogs, cats, and other domesticated animals. For in seeing seals as kin, we also seem to have accomplished a deadly transference of our capabilities onto them. We recognize their prowess as predators and obscure our own, blaming them for fish stocks exhausted by our own overfishing. We bring our efficient machines of profit, greed, and technology to the oceans, and when the hauls peter out, we turn our shame onto the seals and declare them competitors, deserving persecution. In much the same way that we decimated wolf populations, we have spent most of the last three centuries eliminating huge numbers of seals, first through commercial sealing for fur, oil, and ivory, and later to extirpate a marine mammal we mistakenly believed was robbing commercial fishermen. By early in the twentieth century, harbor seals had been nearly exterminated in certain areas, but there was no noticeable effect on fish catches, according to marine scientists and historians.

Reversing the devastation to seal species was a long time coming, and the species' losses—particularly among harbor seals—might well have been irrevocable without the lifting of the bounty on seals in 1962. Now, in Massachusetts, after nearly a century of being stalked by bounty-hunting humans, seals finally appear to have achieved a genuine comeback. And while it is not commonplace, or, as it turns out, entirely safe or wise, to greet seals at extremely close range (even on Monomoy), these marine mammals are frequent, year-round visitors to the shorelines of Cape Cod and nearby islands.

Proof of their success is particularly evident along Cape Cod shores during fall and winter as the bobbing heads of migrating seals dot the waters while the mammals make their way south from Canada. From late fall until early spring seals can be glimpsed from Monomoy and mainland Chatham down to Wellfleet and Provincetown at the Cape's tip, and they are routinely spotted farther north along the Massachusetts coast, at Plymouth, Duxbury, the outer islands of Boston Harbor, and the small islands off Gloucester. Many, according to published reports from the Center for Coastal Studies, in Provincetown, stay all winter.

The dead of winter is actually the peak time to see seals along the New England coast, and isolated beaches along the Outer Cape and Monomoy offer ideal places—perfect for the marine mammals to haul out and sun, and preferable for people wanting to catch sight of pods of young seals or herds of adults. About 75 percent of wintering seals in New England opt to haul out along Cape beaches and Nantucket. And though they are more likely to flee to the water than sit still on land for close observation by humans, it is increasingly common for rugged beachcombers who walk the bitter, blustery outer shore to see the observant eyes and bottling heads of seals—land mammals watching sea mammals watching them—in the surf just offshore.

In fact, seals can be spotted on almost any beach along Cape Cod, and marine and wildlife biologists have witnessed firsthand the recovery of harbor and gray seal populations since 1972, when the federal

Marine Mammal Protection Act was passed. That legislation, plus the reversal of the state bounty on harbor seals a decade earlier, helped to restore seal populations to New England waters. No one knows exactly how many now live year-round along the Massachusetts coastline, but it is estimated that five thousand or more live in the waters off Cape Cod, Monomoy, Nantucket, Martha's Vineyard, and the other small islands nearby. As many as six hundred or seven hundred gray seals alone inhabit the shorelines of Chatham; and by the late 1980s the populations of harbor seals living on the Cape Cod coast had reached several thousand, according to the Center for Coastal Studies. Today, the largest haul-out site in the eastern United States is on Monomoy, where upwards of two thousand seals may be seen when conditions are favorable—that is, when it is more comfortable for seals to be sunning than swimming. As a result, seal-watching boat tours to Monomoy continue, weather permitting, even in winter. Harbor and gray seals frequent the sandy sprawl of Monomoy's outer beaches; and grays, particularly, tend to stay all year, according to reports from the U.S. Fish and Wildlife Service. Occasionally, two other species of seals, harp and hooded, can also be observed.

But seals' acceptance of the presence of humans—and our hangers-on, the dogs—varies from species to species and from one individual to another within herds. Some seals within a species can be very tolerant, while others may be more withdrawn, even belligerent toward an approaching human. Seals do remain alert, however, to action along the shorelines they inhabit. They are fascinated by dogs, and Massachusetts Audubon's Prescott theorizes that the marine mammals see both humans and dogs as beach-bound variations of themselves—

land seals, maybe, only with upright postures. It does not seem such an outlandish guess when one considers that seals themselves, as seen through our eyes, resemble other animals, too. As a group, they are at times offhandedly pegged as "sea dogs," presumably because, when swimming, they appear to have the same sleek, round heads and upraised muzzles of canine water breeds. Moreover, across species of pinnipeds, the young are called "pups."

The resemblance of seals to other mammals goes further than that. The Latin name for the harbor seal—*Phoca vitulina concolor*—translates loosely in English to "sea calf" or "sea dog," for example. This seal, seen in profile, resembles a cocker spaniel with a slightly upturned nose. Likewise, gray seals, when photographed, can appear to be akin to barnyard animals—occasionally they look like sheep, though they are routinely described by naturalists and marine biologists as "horse-head seals" because their heads are massive and sport a nose that is more like a horse's muzzle than a dog's snout. That detail, in fact, is a key to distinguishing between gray and harbor seals.

Seals are agile, powerful swimmers, and as long as there's plenty of food in the ocean, many species ought to be able to flourish just about anywhere. They eat everything from skate, sand eels, and sand dabs to flounder, bluefish, and striped bass. If they run into scant pickings, unable to find enough of their favored fare to satisfy their appetites, they even will take ducks and black guillemots. And in a pinch, they will even fast, occasionally for long periods of time, until prey becomes available. Males of certain species, for example, may fast as long as three months simply to maintain their territory and position in the dominance hierarchy.

Harbor seals far outnumber any other species visible in New England coastal waters, according to Prescott, who, along with other Audubon naturalists and marine biologists, is participating in a federal seal census. Of the harbor seals observed by researchers on the Cape, most tend to be juveniles of less than seven years of age, whereas grays are almost exclusively adults. Furthermore, seal species that once birthed young here but fled when humans arrived in great numbers, are now coming back. For many years, they were pupping only much farther north along rocky outcroppings and beaches such as Sable Island, Nova Scotia, in Canada. But now they have extended their ranges south to include Cape Cod and Nantucket, and in some cases, have reached as far south as Virginia and Florida. In recent years as many as ten gray seal pups were born annually on South Monomoy Island in the dunes facing the Atlantic or on the upper shore.

Mortality rates among the young seals are high, and as many as half die within the first year of life. Over the long haul, seals are still threatened by marine pollution, habitat destruction, or drowning from being trapped in fishing nets. The common denominator in all these perils is the human hand, even the well-meant gestures of humans who happen upon a beached seal and want to help—or, as in my own case, to dally for an afternoon. Most seals, most of the time, don't need to be saved, naturalists and wildlife managers say. Their only need is to be left alone, certainly for twenty-four hours at least. By then, many will choose to leave and will find a way—of their own accord. "Wait . . . and see what happens," Prescott advises, adding a word of caution: "Don't get too close. They will bite."

The Deer Run

The most fundamental attribute of living things is that they can repair themselves.

—Richard Goss, *Deer Antlers*

MY DAYS BEGAN AND ENDED with the white-tailed deer.

Each morning, when the light awakened me to the island, I would pull myself out of my sleeping bag, unsnarling the muslin liner that had become twisted around my legs during the night, and crawl out of the dome tent to check on the weather. I'd lean out of the second-floor windows and throw open the wooden shutters that had slapped closed during a windy night and stare out at the sky and the dawning landscape. From above, even at first light, I'd make my own meteorological calculations from the feel of the damp, foggy, air. Before 7 A.M., the island was almost always shrouded in misty clouds, easing the stark terrain into a vista fluid and gentle as light sleep.

It was just after 5 A.M. on a humid, late-July dawn when the deer caught me unawares. I had been startled awake by the sounds of the great black-backed and herring gulls going at it in the dunes to the east of the lighthouse keeper's cottage. On this morning, like so many others, I was reminded that even on an island lacking human company,

I was hardly alone. I was revisiting my old animal companions. With the birds and animals all around me, my sense of a personal schedule or a private territory slipped away bit by bit, while the barn swallows and song sparrows calculated the dawns and the deer, the oncoming nights. The very season seemed in the creatures' sway, as though the flyers were dragging away the curtain of darkness on their light wings each morning, and the runners towing it back each nightfall.

On this particular morning, long before many people on the mainland would be stirring, I was roused enough to hear that an animal was moving about below on the deck outside the barred front door, scavenging for whatever might be left of the scraps from the previous night's dinner. It had become my ritual to turn over a portion of my food at each meal to the island around me—a way of acknowledging that the island and its creatures were permitting me to share in the place and thus, in return, deserved a measure of my stores. The animal below stayed only briefly to rifle for bits of bread, raisins, and nuts; but a quarter mile off, over the dunes, the gulls continued to carry on, riled into alarm by a predator (the island's lone coyote?) or some other intruder (a fisherman putting to shore?).

The racket got me up and outdoors, and only about five minutes later, so simple were the amenities of getting an island day started. I set a pot of coffee to brew on the portable gas stove, and by then the gulls had settled down into their routine, their cries receding into common quarreling. The animal in the brush had wandered away. I pulled the wood plank that served as a rough-hewn barrier-lock from its perch across the back entrance to the cottage, opened the door, and crossed the threshold into the early morning. I peered out onto the deck and beyond, to a sandscape compressed by fog. Beyond Lighthouse Marsh, everything was liquid and light, and the mist allowed no clear distant sightings in the direction of either ocean or sound. But in the shallows of the near marsh, one outline filtered through the fog, bearing a color like no other in the island palette of greens, browns, and ivory. In the low water, a buck browsed in the bayberry at the pond's edge, and beyond him by thirty yards, a doe stood half hidden in the brush.

The male heard me, I thought, and though he almost surely could not have gotten a sharp look at me, with his keen sense of smell he likely had detected my location. He raised his head out of the leaves and water lilies, his ears pricked high, his white tail twitching. Even under these conditions, with the fog as a foil, I couldn't hide. He knew exactly where I was and locked his gaze on me, as I stood on the deck, in the open door leading from his territory to the other, alien, human world I occupied.

I did nothing but watch for a long moment, then dissolved slowly into the cottage to retrieve my binoculars. The buck never glanced away, not even momentarily, but instead followed my movements across the panorama of the cottage, where, behind closed windows, I was watching, too, checking to see if he would disappear into the scrub farther off. But he stayed and stayed, until I emerged again onto the deck and raised binoculars to my eyes to catch the details of his morning activities.

Whether it was my reemergence outside or the detail of something mechanical and black being lifted, like a gun sight, to my eyes, I couldn't say, but at that moment, the face-off ended. The buck reared his head and began calling in a bleating sound, spreading his influence across the dawn.

It was the first time I had ever been reprimanded by a deer, and I was surprised by the sound of scolding and alarm. But I was to hear it again, within half an hour, when the buck made contact with me again. By then the pair had worked their way around to the other side of the pond, this time approaching from the northeast, through the Hudsonia moors and compass grass. Now the male approached within twenty yards and stopped to watch me sitting in a small beach chair, eating my breakfast. Once more he began to bleat. I took it to be a warning—to his mate?—of the potential human danger and a signal to me that I should venture no closer. But I was content anyway, apart from them, wanting only to observe the deer in their island world.

Meanwhile, I kept an eye on the unfolding day. The sky was still rosy in the east, and dew lay heavy on everything. The laundry I had hung outdoors the afternoon before was getting rinsed again in the natural course of things at daybreak. Just beyond the deck, the drapes of hundreds of webs and filaments of orb-weaving spiders laced the low-lying vegetation, disclosing an intricate, woven world of gossamer traps for insects. A prairie warbler flitted into the *Rosa rugosa* bushes and raised a thin song that was almost instantly overpowered by a yellow warbler tuning in the bayberry ten feet away.

The morning proceeded in the usual fashion of island life in midsummer—tree swallows in acrobatic flight near the pond, meadow voles rustling in the poison ivy off the deck, flies droning on all afternoon inside the cottage and out.

At midday, I hauled a small table and chair onto the deck and wrote all afternoon and into the evening in the sun, pounding away on an old Royal manual that had been resurrected by a typewriter repairman on the mainland. He had sold me the old machine from his stock of used models and had given it a quick tune-up. I had picked it up just before leaving the mainland. When I cracked open the case that afternoon, I discovered a message typed on a scrap of paper in the carriage: "Have a good time camping. Jim."

And so I was.

I finished writing for the day, my copy illuminated by the glow of a propane lantern and the last lights of the sunset. I put a small pot of water on the Coleman stove to boil, tossed in a fistful of pasta, and tore up some shreds of lettuce that had survived the heat without refrigeration. For dessert I sliced up a couple of peaches warmed and softened by the sun and grabbed two oatmeal cookies to go with lemon tea. The meal took all of twenty minutes, start to finish, and I was weighing the possibility of bedding down early for the night when the deer rose out of the marsh and moved nearer, quiet and soft, and now, undisturbed.

We were getting used to each other.

We each had our territory—and safety there.

In the stillness of the island dusk, as the deer lay down like big dogs in the grass by the marsh, I could see our brief and common fate, that we shared the struggle to survive and a sanctuary for rest.

I TEND TO THINK OF MONOMOY as the place that belongs to the birds—clouds of songbirds and shorebirds visiting; gulls supervising everything, everywhere, from their vantage above the sand; harriers lifting themselves up and coasting low to scope the marshes for prey; the godwits and the great horned owl going their own ways—or staying yet a while longer; egrets fishing; plovers hurrying over the near beach; and the silvery terns, skittish along the shores. Because the refuge is swept by such great waves, not just of water but of wings, the island seems to belong, above all, to the birds.

But the white-tailed deer that inhabit Monomoy transform the place. The largest permanent residents of the islands, they are like living symbols of the land itself—visible one moment and vanished the next. I first saw the deer from aboard a skiff on the Sound in late fall in the early 1990s, five of them, midisland, their tails like flags as they vaulted through the grass; and I have seen them since, caught unawares while browsing and scraping in the brush all over the refuge. On South Monomoy during the summer it was not unusual to stumble across the stripped carcass of a young deer, its teeth already worn almost to nubs by the sand-blasted vegetation on which it had, during its short life, fed. Skeletons and scattered bones were riveting clues to how harsh island life could be for a fawn, or even a fully grown buck or doe.

But deer are nothing if not adaptable. Their distant ancestors first appeared about fifty million years ago, but many ultimately faced extinction because of their great size and hulking builds. Two groups of large herbivores persisted—the perissodactyls, encompassing the odd-toed ungulates, such as horses, tapirs, and rhinos; and the artiodactyls, comprising the even-toed ungulates, a large and diversified group of mammals, ranging from hippos and camels to the array of bovids, antelopes, and true deer.

For twenty million years the antlered cervids, or deer family members, have roamed the planet. They are thought to have migrated from Eurasia to North America over the land bridge that once joined Alaska and northeastern Siberia. As glaciers formed, the cervids—which outlasted many of their most terrifying predators, including the saber-toothed tiger and the giant short-faced bear—were pressed southward, arriving in the continental United States about four million years ago. White-tailed deer are a specifically American phenomenon, and today some thirty-eight subspecies inhabit North, Central, and South America. These hoofed herbivores have adapted more successfully than almost any other large mammal to a wide range of conditions and habitats. Part of the reason for their proliferation is that they sustain themselves at the foundation of the food chain, on plants, rather than by preying on other animals. They take a more direct route to food and energy than do the carnivores.

The contemporary subspecies of white-tailed deer vary somewhat from one region to another, a sign of the adaptability that has been a mark of cervid survival over millions of years. Today whitetails can be found from just below the tree line in southern Canada all the way to the rain forests of South America; populations of these deer are in evidence from coast to coast in the United States. They have been so skilled at staying alive and multiplying that the biggest challenge that humans on this continent face with regard to whitetails is "management" of the large numbers of deer to keep them in check and ensure the health of the herds.

Popular lore and some anecdotal evidence suggest that the deer on Monomoy are smaller than the average white-tailed deer of the northeastern and midwestern United States, but wildlife experts are not of one mind on the subject, and no formal studies of the Monomoy herd have been done. On one hand, the notion that the size of individual bucks and does might be, on average, smaller than those in mainland herds makes sense; it is certainly true that the Monomoy herd has limited supplies of quality browse with which to build size or bulk. The island whitetails sustain themselves on a diet of beach grass, woody bayberry, poison ivy, spartina grass, seaweed (such as eel grass), and aquatic vegetation growing around the larger marshes and ponds; but they do not enjoy the species' preferred diet of more succulent grasses, sedges, fruits, and nuts. Given the available territory, it seems remarkable that the deer herd can persist over time, especially considering that the refuge encompasses mostly sandy habitats of rugged but relatively poor-quality flora, high in salt content.

On the other hand, the herd on Monomoy has an advantage that mainland deer do not enjoy—almost complete freedom from predators. Though hunting was popular for decades during the nineteenth and early twentieth centuries on the then-peninsula of Monomoy, it has been outlawed since the 1940s, when the federal government took over the land. Isolated incidents of poaching have continued until fairly recently; but even with the illegal shooting of deer, the stress on the population and the culling of the herd in no way rival the taking of deer on the mainland. In areas with large deer populations, humans are, by far, the most skilled and successful predators, culling hundreds of thousands of deer—mostly bucks—from continental herds during hunting season.

Freedom from predation not only by humans but by other carnivores is one of the key arguments advanced in favor of the general health and heft of the Monomoy herd. A doe in northern Vermont or inland Maine, for example, might elude humans, only to be being taken down by a coyote, bobcat, black bear, or domestic dog. Certainly, if she is well, she is equipped to evade some predators. Biologists and wildlife specialists have clocked white-tailed deer running faster than thirty-five miles per hour and have documented their ability to clear a seven-foot fence from a standstill.

On Monomoy, by comparison, the greatest likelihood is that death will result from starvation, malnutrition, disease, or the overall strain of enduring severe winters without adequate shelter or cover. Almost none of the common predators or problems of mainland deer plague the islands' herd. In addition to their protection by law from people, the deer have been shielded from other animals by the surrounding waters and by Fish and Wildlife management efforts, which led to the removal of some mammals in the 1950s. For nearly a half century, the only significant mammal populations have been deer, muskrats, meadow voles, and mice. In the mid-1990s, however, coyotes—having established themselves on mainland Cape Cod—made their way to Monomoy and posed a potential threat to the small deer herd as well as to a number of nesting and endangered birds. The Fish and Wildlife Service, under federal mandate to manage mammals that pose a threat to avian diversity, was required to remove any coyote pairs attempting to den there to reproduce young. Because the first pair gave every indication of making the island its home turf, Fish and Wildlife officials imported a wildlife specialist from Colorado to eliminate the coyotes. The female of the pair was shot and killed in 1998, and in the spring of 1999, a second female was also killed. Transient male coyotes or solitary individuals will not be hunted, unless they move into nesting territories.

The deer herd cannot multiply beyond the carrying capacity of the land. Since the 1950s, the herd size has been as high as forty-five or fifty, but the more constant number during the last decade has been about twenty-five. Although white-tailed deer on the continent can move into new territory if numbers necessitate expanding their range, on Monomoy, land runs out at the shores. Fishermen and wildlife officials have observed individual deer, presumably those unable to find adequate food or sufficient range, swimming toward Chatham and Orleans, or even making it as far as Nantucket, some fifteen miles away, where they reach the island and fall, exhausted, onto the beach.

A visitor to the refuge is not guaranteed a glimpse of the whitetails, but it is fairly common to see solitary deer or as many as five or six feeding together near the marshes stretching west from the lighthouse toward Nantucket Sound. As is typical of the species in New England, they are most active just after dawn, at midday, around dusk, and a couple of times in the dead of night.

When and where they eat, and how they graze, is determined not by their appetites or how palatable the browse is but rather by what offers the best defense against predators. They remain hidden or bedded down during the hours when coyotes are out hunting, letting camouflage and cover do the work of eluding potential killers. And when they do feed, they come to the task equipped with a digestive system that allows them to stuff themselves quickly and then later, by regurgitating, or ruminating, to complete the chewing and processing of food in safety.

This "eat now, chew later" policy, as one wildlife researcher describes it, is essential for animals that fuel their bodies with vegetation. Deer are ruminants, or cud chewers, with a compound, four-chambered stomach. The first two portions of the stomach are specialized in ruminants and are designed efficiently for the difficult job of breaking down the tough walls of cellulose in plants for absorption into the body. On their own, most mammals cannot do this; it is the job of microorganisms in the digestive tract.

This complex digestive tract is but one of the several adaptations deer have made as part of their long, prolific sojourn on this planet. It pinpoints two critical achievements for survival: a way to make the best use of the most plentiful kind of food in its habitat, and a means to concentrate on food without forgetting the risk of harm from predators.

But the physiology of individual deer as well as the behavior and social organization of herds have had to evolve to enhance their chances for survival. There is no area of life or limb in deer that has not been affected by the drive of life for more life. Every shudder of evolutionary adaptation reflects how inextricably linked the animal is to the constantly changing conditions of its habitat and other creatures that share it. Once primarily a species of sheltering woodlands, deer have spread over much of the world to more open grasslands, marshes, swamps, mountains, and islands. And though they live for much of the year segregated by sex and frequently in small groups, they are social animals.

Just to take a close look at the body of a whitetail is to witness the broad specializations that have earned it such prominence in the struggle for life. It developed long limbs and strong muscles for eluding other animals. Even its feet, which keep the animal perched on two digits, are designed for quick takeoff and the graceful, bounding flight for which deer are so admired. Deer evolved large, flat-topped teeth with interlocking folds of cementum and enamel that create sharp edges for grinding fibrous or woody plants. Additionally, a deer has a flexible tongue, enabling it to browse selectively and to take fibrous or woody plants, if more desirable grasses and leaves are scarce. Its large eyes, keener in capturing movement in daylight or darkness than in discriminating color and form, are set into the near sides of its head to enable a broader field of vision; and its hearing and sense of smell are extremely acute. Its olfactory sense is so sharp and sensitive that a whole range of communications relies on scent, from breeding during the rut to the imprinting of fawns by does.

Deer have a variety of visual and vocal cues to signal alarm, fear, aggression, even encouragement (as in the gentle "mewing" of does to fawns). Depending on mood and message, they will bleat, grunt, snort, or wheeze; a look can express enough to save them from having to expend valuable energy fighting or chasing other deer. Whitetails will spell out their intent through a whole alphabet of aggressive or dominance vocalizations, facial expressions, and body language. Wildlife biologists have identified a number of these signals, including the ear drop, hard look, sidle, rush, snort, strike, and flail, as escalating aggressive warnings about social hierarchy and a means of sorting out which animal is in charge and which must submit. Outright physical battles are relatively rare and tend to be mostly limited to males vying over females.

Deer also leave complex scent messages—not all of which we understand, despite considerable research. We do know that this chemical communication can be transmitted through seven different scent glands in their skin, urine, vaginal secretions, and possibly saliva, as well. Three scent glands are situated on a deer's head, three are on its legs, and one is on the sheath of its penis. A bewildering array of messages is handled with a system of behaviors known as "signposting" or "buck rubs": Full-grown dominant males, again, most commonly during the autumn rut, or breeding season, will score tree branches with their antlers or press against them at a scent gland site. Likewise, they will leave "scrapes," round or oval pawed areas of soil and, sometimes, leaves or other debris on which they urinate; and "licking sticks," branches or twigs that have been nuzzled or licked, for other deer to "read." Whatever else these behaviors and scents signal, they almost certainly relay information about the bucks' identities and social status both to competing bucks and to does.

But the principal symbol of status among bucks is the antler rack—precisely what humans find impressive about these animals. It is striking just for what it is—external bone that grows into a fierce set of weapons. Then there is the almost incomprehensible phenomenon of antler casting, in which bucks shed their racks, and the following spring, with astonishing speed, regrow a new set of velvet-covered antlers, which eventually harden to bone by summer. The process and its product belong uniquely to the deer family, Cervidae, and are unprecedented among mammals. Even horns, which to most people seem similar to antlers, are quite different. They are made of keratin, a substance found in hooves, and they are permanent (except in pronghorn antelope), growing a bit each year from points at the base of the skull.

Shedding and regrowing are triggered by the seasonal changes in the length of daylight. These stimulate the pineal gland to release, or alternatively, to inhibit, melatonin, which in turn affects the level of testosterone, which triggers antlers to grow and, later, to harden.

While they are growing, antlers are living tissue, and the velvety surfaces are sensitive to touch. Early on, they are soft and can be damaged. If a healthy buck's antlers are injured at this stage, he may grow a new single one or set during the same year. Ordinarily, antlers begin to grow in the spring, issuing from a specialized, spongy bone stump, or pedicle, which is essential for antler formation. (Researchers have demonstrated this by surgically grafting pedicle tissue to legs and other parts of deers' bodies, whereupon antlers begin to form.) It takes about a hundred days for a full set of antlers to grow, and during that period—which ends as the shorter days of late summer and early fall begin—the antlers are warm, reaching a temperature as high as that of the buck's body core. The antler lengthens, slowly at first, then more quickly as summer wears on, growing from the tip at the rate of about a quarter inch per day. The bone calcifies from the base, and as the summer wears on, the velvet dies and the entire antler hardens. There is a corresponding and crucial dieback of the tissue at the point where the pedicle meets the antler. If that tissue does not die, the antlers will not be shed, and the tissue going down the pedicle and into the skull will die, with fatal consequences for the buck. Ordinarily, though, the buck retains his hardened antlers throughout the fall rut and into winter, while his testosterone levels are high. Then, as winter gives way to early spring, the antlers are shed, and as longer daylight returns, the process starts again.

Growing and casting antlers each year requires a lot of energy, and to biologists the process seems wasteful because no one knows exactly what function the antlers serve. Theories range from the idea that antlers developed by some twist of evolutionary caprice to the notion that they developed side by side with behavior and social organization. It is evident that needed or not, antlers were powerful tools for communication of male rank as well as weapons of defense and aggression in a range of circumstances with other deer or against predators. Still, in many cases, the size of the rack seems to outstrip its purpose, representing what Richard Goss, an authority on antler regeneration, function, and evolution, describes as "an extravagance of nature rivaled only by such other biological luxuries as flowers, butterfly wings, and peacock tails. The antlers of deer are so improbable," he contends, "that if they had not evolved in the first place, they would never even have been conceived in the wildest fantasies of the most imaginative biologists."

THE WHITETAILS MOVE with ethereal grace within the islands' shores. Like the ghost ships of marine folklore that run aground night after eternal night on the shoals facing the open Atlantic, the Monomoy deer appear as specters traveling corridors of their own devising, from pond to marsh, from shore to inland moor. Whether on the island or the continent, white-tailed deer remain among the most superior examples of animal beauty, majesty, strength, and grace in all of nature. Of the popular wild-animal icons of contemporary Western culture—including whales, dolphins, seals, penguins, and wolves—deer are said to be the most adored, despite their status as pests in the flower and vegetable gardens of suburbia. Part of their appeal derives from the assurance that they pose no threat to humans and that almost everyone, whether a rural or urban dweller, has seen a deer firsthand.

And then there is the matter of the eyes.

And the antlers.

And the hobbling, vulnerable young.

FAWNS ARE BORN IN SPRING, well after the most intense winter storms. A pregnant doe announces the imminent birth—likely twins, if she is in good health—by licking her flanks or pacing with her tail raised. Her udders will swell shortly before the fawns are birthed. Once the young have been born, she will sequester them from other members of the herd, as well as from each other, so that, in seclusion, she can imprint them to her and "memorize" their particular scent.

At birth, fawns are wobbly and awkward, but they can walk within minutes, and after a half hour they are able to be moved to separate "visitation" sites, which might be as much as five hundred feet apart. There they will remain, checked on, nursed, and groomed by their mother three times a day for two and a half to four weeks, or slightly longer, if needed. When they are between eighteen and thirty-two days old, they will begin bedding down together, but until they are older and stronger, their best chance at survival is hiding. A fawn's spotted fur is the perfect camouflage for the dappled light of a thicket in summer or the brush edging a marsh or pond. Of all the seasons, summer represents the closest thing to rest and ease that a deer can enjoy in the wild, and a lighthearted mood erupts early in newborn fawns. At a week old, some will break out in play, running, jumping, and kicking—but always within sight of the doe. Even at this, nature's hidden agenda is a primer in survival, for the very behaviors fawns express in play they may well need later for defense against predators or competitors—a quick flash of the tail and swift evasion and flight.

Reproductive rhythms—whether the event is estrus, rutting, or bearing young—are tied to the length of daylight, so that fawns are born under conditions most compatible with survival. In the northern United States, the first signs of spring correspond with the running of maple sap, but because cold and snow can freeze the landscape even in late April and on rare occasions in May, the spring season may best be described as ephemeral, more a promise than a certainty until mid-May. On Monomoy, the brief spring season arrives late and lasts well into June; even on Memorial Day, conditions on the islands frequently are damp and chilly, and the waters of the Sound are uncomfortably cold for swimming.

Time for birds and animals is a measure of real essentials of the life cycle. For whitetails, the seasons are governed by physical necessities and reproductive dictates: spring for giving birth; summer for nurturing and raising young, as well as for does and bucks alike to build their bodies in preparation for the rut; fall for mating; and winter for just getting by.

Enduring the winter is the harshest of the earliest challenges for young whitetails, particularly on a narrow, sand-spit island like Monomoy, where there is scant shelter for the herd's protection. On the mainland, deer can seek the cover of conifer stands for woody browse to keep them alive as well as to avoid the blast of winter storms on open meadows, marshes, and plains. Monomoy is itself that protection for the mainland: a barrier island absorbing the devastation of blizzards, high winds, and furious seas. It is far from an ideal place for foraging animals to take shelter. As the March 1958 and February 1978 breaches of the islands proved, safety can be swept away—literally—in a single night.

Still, like the island itself, they endure, fleeting as the light and the shore, lasting as a miracle.

THE FIRST DEER, A BUCK, remained aloof, standing on a hillock of sand a hundred feet off, staring at me staring at him. The second, a doe, stood for only a second, six feet away, on the other side of a stripped stand of *Rosa rugosa,* then bolted into the bayberry and low brush.

They were my second and third undeserved blessings of the day.

It was late September, and I thought I knew enough of island time to predict that most of the shorebirds would be gone, so it was a surprise to see the sanderlings pacing ahead of me at the edge of the surf, darting up and down the shoreline with the waves. The day was bleached by sun, and the light was breaking off the crests like foam. From the Nantucket Sound side of South Monomoy, I could see across to the ocean, the white waves tossing bits of themselves out of the sea like spirits ascending into thin air from the insubstantial body of the silver water. Empty whelks, months ago carried inland by gulls, lay strewn across the expanse of Hospital Pond at the island's north end, and the dried beach grass etched whorls and coils as if signing off on something more complete and essential than the end of a season.

Then, the buck was there, up ahead and inland, intent but not alarmed, it seemed, as he watched me, stopped on the beach at the sight of him. For a full minute we regarded each other, neither of us moving muscles other than eyes, ears, and nose—the stillest active senses by which to measure the possibilities. Then he turned a broad shoulder to me, leapt off the mound, and disappeared.

I was still held by the power of the male's look, mute and wild, when the doe lifted her head over the clump of bushes at the edge of the pond where, in summer, the snowy egrets and black-crowned night herons congregate. She seemed as startled as I to be so close, one to another, two potential enemies under ordinary circumstances, especially on the mainland. Then she flashed the white underside of her tail and bounded off.

Walking the shoreline, poking in the ruins of shells and seaweed, hoping to uncover a bone, I felt a shock of recognition when I came unexpectedly on the deer, and wondered later what it was I thought I caught in those huge, unblinking eyes, wearing me down to silence and submission. Attention? Innocence? Alarm? Or some other uncommon sense of things, linked not to forethought or a language of syllables strung into words but to the dictates of blood and bone?

I know something of how they exist, the bucks keeping to themselves except during the rut, when all the usual bets are off about males and females living in segregated groups or even in solitude. I understand that there must be thick-limbed rivalries and the challenge of rack on rack to determine which microscopic gene-power will win the day and the doe, to be passed on to the next generation.

I have seen across the chasm that stretches from my human intelligence to the wide-eyed way the whitetail knows, felt the wonder of that split-second stare that brings neither discernment nor fear, when only a presence, an Other, floods the retina. I could drown in the eyes of a doe, unblinking and full of the dark earth: drown, and surface again, miraculously, beyond hope or expectation, and lay my heavy animal self down against the breast of a dune on a mat of crushed conifers, laden with dreams and desire. For safety's sake, I'd keep my snout keyed to the wind and learn to survive with the scents the winds carry. To conceal me from the unnatural world, I might even find my own natural disguise—cloven feet, velvet antlers, and pure hunger—belong once more to the herd and, with my kind, learn to run with the power of necessity, the energy of life.

THE LAST FULL DAY of my third summer stay on Monomoy, I passed the time following the pathways sculpted by the white-tailed deer. I had decided to walk down to the southern end of South Monomoy and look out over the riptides and shoals that for four hundred years have made passage around the point a risky, and sometimes deadly, enterprise. I wanted to devote my final-day walk to the part of the island I knew least—the ponds. Nestled in the dunes near the point lie Big and Little Station Ponds, where the waterfowl seek haven, particularly during fall migration. At the peak of the autumn migration, from mid-August to mid-September, thousands of ducks, shorebirds, and seabirds stop over, crowding into the waters of Big Station Pond for rest and feeding before heading on to destination points thousands of miles south of the refuge.

It was early August, and I was running a little ahead of the most intense press of the fall migration, though that restlessness that I associate with the long flights already was palpable in the air. As I made my way, at last, around the far side of Big Station Pond, I saw that the dawdlers were nothing spectacular, just the comical, routine cormorants, scores of them in the shallow waters, and a handful of sanderlings probing the shore. Even so, there was something primordial—almost otherworldly—about this habitat I had never visited before. In one marshy spot off Big Station Pond, the drought had isolated a small pool from the main body of water, creating a miniature pond perhaps four inches deep. There, three small bass hid, their dark colors camouflaging them even in the meager dank waters evaporating all around them. It occurred to me to try to catch them in my hands and carry them the twenty feet that separated their little world from the big pond; but in the end, I shrugged the idea off, letting nature manage its own harsh formula and leaving myself out of the bargain.

I had navigated my way to the marsh only by the trail blazing of the deer, who had broken a pathway through the dense thickets of bayberry and poison ivy growing chest-high on every side of the pond. In the wet sand, I could see their cloven tracks, and after an hour or so of wandering in the brush, I was able to discern almost immediately where a slight depression in the grass or an insignificant parting in the bush signaled a trail crafted by the elegant, powerful body of a deer.

It gave me an odd sense of satisfaction—contentment, even—to be following in the footsteps of a deer, as if I was assuming my proper, respectful position in the island's chain of being. The deer seem to have achieved a higher state of evolution than I have accomplished, knowing as they do what their job is, survival and reproduction, and finding completion just in doing it. In an irrational moment of magical thinking, I pretended that following the deer could perhaps lead me to that same destination: a sense of living with purpose, with no brooding or deliberation to slow me down.

Instead, the deer trails led to the water and out again onto the dunes, along sandy shorelines ornamented with downy feathers of gull chicks and marsh skullcap just beginning to bloom. I gave up on the looming questions and relinquished my hopes for big answers, settling for collecting some pebbles and bits of driftwood as keepsakes of the day and the place. Finally, I waded through the shallows from one shore to another, reaching the far side with soaked shoes, the legs of my jeans clinging and heavy as hocks, my heart light as first feathers, and unexpectedly filled with a simple, perfect peace.

I hadn't seen a deer all day long, anywhere but in my mind. I pictured them moving quietly ahead of me through the bayberry, or heaving their bodies down into the compass grass, knowing where to seek safe respite—and how to hold still right there. I envisioned all they might teach me about this island and its enigmas, their secrets—something elusive about beauty and innocence and grace.

That evening, seven white-tailed deer came to browse again in Lighthouse Marsh. I watched them for nearly an hour, memorizing their movements and their stillness to recall later, in another place, when I longed for their company but could not get close.

I left the next morning, hauling load after load of gear out to Nantucket Sound, hollering at the gulls hollering at me. I was intent mainly on my burden as I made my final portage for the trip back to the mainland, my head down, my back bent under the weight of the pack, a stick held over my head to discourage the gulls from their distracting, worrisome maneuvers.

And suddenly, as if out of the thin, bright summer air, there they were again: three deer, the nearest one only twenty feet away, the most distant not more than fifty feet off. For a moment I forgot everything except those eyes on me—and the feeling that there was no harm anywhere in the landscape. That assurance was sufficient promise for me, comfort enough to keep me on track, back to the human world over there, where the sea waters cease and unquiet civilization thrums.

Epilogue

When you walk across the fields with your mind pure and holy, then from all the stones, and all the growing things, and all animals, the sparks of their soul come out and cling to you, and then they are purified and become a holy fire in you.

—Hasidic saying

WHO CAN SPEAK the violence of nature, and who can chart its splendid peace?

The waters that spewed us—a family forming, genus not yet entire, nor species ever finished—rise up again and again against us, daring to pull us back into the primordial deep. Waterspout, mud slide, hurricane, drought, ice—a hundred pestilences in the wind directed to balance earth and sky. Once more, the skies clear and the land opens before us, hope defining the near frontier and yearning the distant horizon.

The swallow flies, nest to roost. The owl, in silence yet, with soundless flight, haunts field and thicket; and its prey takes whatever cover comes—empty burrow, fallen leaf, rotting log. From saltwaters the sea lion bull hauls himself, seeking air and frightful contest, harem and mates, or death. All for life, this ascent and decline, more life always, multiplied beyond number, driven to limits only death and desire can impose.

Look to the heavens, the works there, the moon and the stars that have been established, the ordering of the infinite. What are these creatures below, beings of bone, air, and blood, who have put all things beneath their feet, all the beasts of the field, the birds of the air, the fish of the sea, and all that passes along the paths of the sea. Erect, they walk, terrible beauty in their bearing as they move, these carriers of fire and flickering light. But where is the way to the dwelling of light, and where is the place of darkness? Declare all that is known.

Monomoy.

IN THOSE ISLAND YEARS, it came to me that the language belonged to the land, and to the creatures of the land, beyond us all. Disconnection would unite us, I saw then, and death would publicize life. My mother died first during that decade, then later, my father. And it was at that ritual of grieving—as we buried the empty shell of the body, the secrets, the shame—that I saw I might be restored by relation, my siblings kin, too, in the family of things. And though we hardly knew what the great drift of years apart had meant, it was a beginning again.

One evening just after my siblings and I had laid the body of our father to rest, I took a walk, alone for solace, in the twilight at the bog. I had flown across the days and nights to the Midwest and back east again—the longest journey of my brief life. Now it was almost fully dark as I left the car at the edge of the woods a half mile from home and let the dog lope out ahead into the dusk deepening all around us. I set out, wanting nothing, expecting less, giving up the emptiness inside me to the open space over the red and sodden fields. I heard my shoes whine softly in the sand, listened to the loud, indifferent wind wrap around my parka, felt my own heart beating hard, and heard, far off, the first peeper of the season cry, solo, in the dark—all the sounds of spring going on as if there were no ghosts in the landscape, nothing but new life and second chances.

I was walking the short corridor of unpaved road lined with trees when the birds interrupted my grieving. Two cardinals—a male and a female—broke out of the thicket, flying fast and low. At first I thought they must be bats, given the late hour and the darkness, but they flew so close to my face that I could see, even in the faded light, the unmistakable form and the crimson color of the male's feathers.

Cardinals flying at night, I thought, the impression as heavy as clay in my mind. They had been my mother's favorite birds—like the jays I loved, an ornament of color in the dull urban landscape of Chicago, where I had been raised. As a child I had learned to mimic their call, whistling to them for blocks, moving through the territory of one pair after another on my way home from school.

Cardinals in the dark, I marveled softly, as the forms veered away, out of sight—my mother dead nearly four years, my father, five days. I kept moving, telling myself I had witnessed something important, telling myself stories, edging the bog slowly because I had no better place to go with my sorrow or my love than to this old, familiar circle.

Tonight I know mercy, I thought without thinking at all—and felt something small, inside me, break apart like a dry twig split underfoot.

Tonight my world is lighter by one. By two.

And overhead, a pair of Canada geese moved in on the marsh, complaining all the while, as they heaved their great bodies to the surface of the still and carefree waters. The new moon, like an eye barely open, hung above the stand of trees, and the stars came out. Life goes on and on, I thought, deriving no comfort from the fact, but sensing a hope as palpable and mere as the season stirring in the terrain. This is neither bad news, nor good, I knew then, suddenly feeling my part as definite as the oldest oak in the marsh, as elemental as the earth. This is simply the lay of the land.

Bibliography

General

Amos, William H., and Stephen H. Amos. *Atlantic and Gulf Coasts.* New York: Alfred A. Knopf, 1988.

Bailey, Wallace, and Priscilla Bailey. *Monomoy Wilderness.* South Wellfleet, Mass.: Massachusetts Audubon Society, 1972.

Cairn, North. "Monomoy: Paradise Found." *Cape Cod Times* (series), February 1994. Also miscellaneous news and feature articles, June 1998 through August 1999.

Carson, Rachel. *The Edge of the Sea.* Boston: Houghton Mifflin, 1979.

Gosner, Kenneth L. *A Field Guide to the Atlantic Seashore.* Boston: Houghton Mifflin, 1978.

Hay, John, and Peter Farb. *The Atlantic Shore: Human and Natural History from Long Island to Labrador.* Orleans, Mass.: Parnassus Imprints, 1966.

Keatts, Henry. *Beachcomber's Guide from Cape Cod to Cape Hatteras.* Houston: Gulf Publishing, 1995.

Ludlum, David. *The Country Journal New England Weather Book.* Boston: Houghton Mifflin, 1976.

Newcomb, Lawrence. *Newcomb's Wildflower Guide.* Boston: Little, Brown, 1977.

The Penguin Book of the Natural World. Baltimore: Penguin Books, 1976.

Riley, Laura, and William Riley. *Guide to the National Wildlife Refuges.* New York: Macmillan, 1979.

U.S. Department of the Interior. Fish and Wildlife Service, Region 5. *Environmental Assessment Master Plan for Monomoy National Wildlife Refuge.* Newton Corner, Mass., 1988.

Ward, Nathalie. *Stellwagen Bank: A Guide to the Whales, Sea Birds and Marine Life of the Stellwagen Bank National Marine Sanctuary.* Provincetown, Mass.: Center for Coastal Studies, 1995.

Archaeology and History

Bachand, Robert G. *Northeast Lights: Lighthouses and Lightships, Rhode Island to Cape May, New Jersey.* Norwalk, Conn.: Sea Sports Publications, 1989.

Braun, Esther K., and David P. Braun. *The First Peoples of the Northeast.* Lincoln, Mass.: Moccasin Hill Press, 1994.

Carpenter, Delores Bird. *Early Encounters: Native Americans and Europeans in New England: From the Papers of W. Sears Nickerson.* East Lansing: Michigan State University Press, 1994.

Dalton, J. W. *The Life Savers of Cape Cod.* Chatham, Mass.: Chatham Press, 1902.

Finch, Robert. *Cape Cod: Its Natural and Cultural History.* Washington, D.C.: National Park Service.

Floherty, John J. *Sentries of the Sea.* New York: J. B. Lippincott, 1942.

Gill, Crispin. *Mayflower Remembered: A History of the Plymouth Pilgrims.* New York: Taplinger, 1970.

Johnson, Robert Erwin. *Guardians of the Sea: History of the United States Coast Guard, 1915 to the Present.* Annapolis: Naval Institute Press, 1987.

Johnson, Steven F. *Ninnuock (The People): The Algonkian People of New England.* Marlborough, Mass.: Bliss Publishing, 1995.

Jones, Maro Beath. "Monomoy Light Adventure" and unpublished diaries. Private collection of Emily Aasted, Santa Barbara, Calif.

Kemp, Peter R., ed. *Oxford Companion to Ships and the Sea.* Oxford: Oxford University Press, 1988.

Lombard, Asa Cobb Paine, Jr. *East of Cape Cod.* Cuttyhunk Island, Mass.: Reynolds DeWalt Printing, 1976.

Nickerson, W. Sears. *Land Ho!—1620: A Seaman's Story of the Mayflower, Her Construction, Her Navigation and Her First Landfall.* Cambridge, Mass.: Houghton Mifflin, 1931.

Putnam, George R. *Lighthouses and Lightships of the United States.* Cambridge, Mass.: Houghton Mifflin, 1933.

Quinn, David B., and Alison M. Quinn, eds. *The English New England Voyages, 1602–1608.* London: Hakluyt Society, 1983.

Russell, Howard S. *Indian New England before the* Mayflower. Hanover, N.H.: University Press of New England, 1980.

Shanks, Ralph, Wick York, and Lisa Woo Shanks, eds. *The U.S.*

Life-Saving Service: Heroes, Rescues and Architecture of the Early Coast Guard. Petaluna, Calif.: Costano Books, 1998.

Thoreau, Henry D. *Cape Cod.* 1865; reprint, Princeton: Princeton University Press, 1988.

Smith, William C. *A History of Chatham, Massachusetts.* Chatham, Mass.: Chatham Historical Society, 1971.

Steger, Will, and Jon Bowermaster. *Crossing Antarctica.* New York: Dell Publishing, 1991.

Thompson, Frederic L. *The Lightships of Cape Cod.* Portland, Maine: Congress Square Press, 1983.

Waldman, Carl. *Atlas of the North American Indian.* New York: Facts on File, 1985.

Birds and Migration

Alerstam, Thomas. *Bird Migration.* Translated by David A. Christie. Cambridge: Cambridge University Press, 1990.

Dingle, Hugh. *Migration: The Biology of Life on the Move.* New York: Oxford University Press, 1996.

Dorst, Jean. *The Migrations of Birds.* Boston: Houghton Mifflin, 1962.

Elphick, Jonathan, ed. *The Atlas of Bird Migration: Tracing the Great Journeys of the World's Birds.* New York: Random House, 1995.

Griffin, Donald R. *Bird Migration.* New York: Doubleday, 1964.

Hill, Norman P. *The Birds of Cape Cod, Massachusetts.* New York: William Morrow, 1965.

Koenig, Otto. *Family Life of Birds.* New York: McGraw-Hill, 1971.

Matthews, Geoffrey V. T. *Bird Navigation.* London: Cambridge University Press, 1968.

Megyesi, Jennifer Lynn. "1997 Field Season Report, Management for Avian Nesting Diversity on the Northern End of South Monomoy Island," *Environment Cape Cod* 1, no. 3.

National Geographic Society. *The Wonder of Birds.* Washington, D.C.: National Geographic Society, 1983.

Pasquier, Roger F. *Watching Birds: An Introduction to Ornithology.* Boston: Houghton Mifflin, 1997.

Peterson, Roger Tory. *A Field Guide to the Birds: Eastern and Central North America.* Boston: Houghton Mifflin, 1980.

———. Introduction to *The Birds of Cape Cod, Massachusetts,* by Norman P. Hill. New York: William Morrow, 1965.

Short, Lester L. *The Lives of Birds.* New York: Henry Holt, 1993.

Stokes, David W. *A Guide to the Behavior of Common Birds.* Boston: Little, Brown, 1979.

Terborgh, John. *Where Have All the Birds Gone?* Princeton: Princeton University Press, 1989.

U.S. Department of the Interior. Fish and Wildlife Service, Region 5. *Atlantic Coast Piping Plover Recovery Plan.* Newton Corner, Mass., 1988.

Veit, Richard R. *Birds of Massachusetts.* Lincoln, Mass.: Massachusetts Audubon Society, 1993.

Waterman, Talbot H. *Animal Navigation.* New York: Scientific American Library, 1989.

Whitaker, John O., Jr. *Audubon Society Field Guide to North American Birds.* New York: Alfred A. Knopf, 1980.

Geology, Oceanography, and Coastal Processes

Bush, David M., Orrin H. Pilkey, Jr., and William J. Neal. *Living by the Rules of the Sea.* Durham, N.C.: Duke University Press, 1996.

Chamberlain, Barbara Blau. *These Fragile Outposts: A Geological Look at Cape Cod, Martha's Vineyard and Nantucket.* Yarmouthport, Mass.: Parnassus Imprints, 1981.

Dean, Cornelia. *Against the Tide: The Battle for America's Beaches.* New York: Columbia University Press, 1999.

Fox, William T. *At the Sea's Edge: An Introduction to Coastal Oceanography for the Amateur Naturalist.* Englewood Cliffs, N.J.: Prentice-Hall, 1983.

Goldsmith, Victor. "Coastal Processes in a Barrier Beach Complex and Adjacent Ocean Floor: Monomoy Island—Nauset Spit, Cape Cod, Massachusetts." Ph.D. diss., University of Massachusetts, 1972.

Hoel, Michael L. *Land's Edge: A Natural History of Barrier Beaches from Maine to North Carolina.* Newbury, Mass.: Little Book, 1986.

Kaufman, Wallace, and Orrin H. Pilkey, Jr. *The Beaches Are Moving: The Drowning of America's Shoreline.* Durham, N.C.: Duke University Press, 1983.

Kopper, Philip. *The Wild Edge: Life and Lore of the Great Atlantic Beaches.* New York: Penguin Books, 1979.

Leatherman, Stephen P. *Barrier Island Handbook.* College Park, Md.: Coastal Publications Series, Laboratory for Coastal Research, University of Maryland, 1988.

———. *Cape Cod Field Trips: From Yesterday's Glaciers to Today's Beaches.* College Park, Md.: Coastal Publications Series, Laboratory for Coastal Research, University of Maryland, 1988.

Oldale, Robert N. *Cape Cod and the Islands: The Geologic Story.* East Orleans, Mass.: Parnassus Imprints, 1992.

Mammals

Bonner, W. Nigel. *Seals and Man.* Seattle: University of Washington Press, 1982.

Canton, John Dean. *Antelope and Deer of America.* New York: Forest and Stream, 1877.

Coffey, David J. *Dolphins, Whales and Porpoises.* New York: Macmillan, 1977.

Goss, Richard J. *Deer Antlers: Regeneration, Function, and Evolution.* New York: Academic Press, 1983.

Katona, Steven K., Valerie Rough, and David T. Richardson. *A Field Guide to Whales, Porpoises and Seals from Cape Cod to Newfoundland.* Washington: Smithsonian Institution Press, 1993.

Maas, David R. *North American Game Animals.* Minnetonka, Minn.: Cowles Creative Publishing, 1997.

Newsom, William Monypeny. *Whitetailed Deer.* New York: Charles Scribner's Sons, 1926.

Ozoga, John J. *Seasons of the Whitetail: Books One through Four (Whitetail Autumn, Winter, Spring, Summer).* Minocqua, Wis.: Willow Creek Press, 1997.

Prescott, Robert. Lecture on seals of Cape Cod, presented at Snow Library, Orleans, Mass., November 1998.

Putnam, Rory. *The Natural History of Deer.* Ithaca, N.Y.: Cornell University Press, 1988.

Reidman, Marianne. *The Pinnipeds: Seals, Sea Lions, and Walruses.* Berkeley: University of California Press, 1990.

Mythology, Spirituality, and Folklore

Anderson, Lorraine, ed. *Sisters of the Earth: Women's Prose and Poetry about Nature.* New York: Vintage Books, 1991.

Arrowsmith, Nancy, with George Morse. *A Field Guide to the Little People.* New York: Hill and Wang, 1977.

Frazer, Sir James George. *The Golden Bough: A Study in Magic and Religion.* Reprint, New York: Macmillan, 1951.

Jamal, Michele. *Deerdancer: The Shapeshifter Archetype in Story and in Trance.* New York: Arkana/Penguin, 1995.

Manes, Christopher. *Other Creations: Rediscovering the Spirituality of Animals.* New York: Doubleday, 1997.

Palsson, Gisli. "The Idea of Fish: Land and Sea in the Icelandic World-view." In *Signifying Animals: Human Meaning in the Natural World,* edited by Roy G. Willis. London: Unwin Hyman, 1990.

Perry, Whitall N. *A Treasury of Traditional Wisdom.* San Francisco: Harper & Row, 1971.

Smart, Ninian, and Richard D. Hecht, eds. *Sacred Texts of the World: A Universal Anthology,* New York: Crossroad, 1982.